Naturopathic Treatment of Emotional Illness

C.P. Negri, OMD, NMD
with
Randall Gibson, M.Ed., LMT

Naturopathic Treatment of Emotional Illness

By C.P. Negri, OMD, NMD, with Randall Gibson, M.Ed., LMT
Illustrations by C.P. Negri
First edition

ISBN: 978-0-9819884-3-6

Care has been taken to confirm the accuracy of the information presented in this work. While some data and procedures are generally accepted practices in natural medicine, many are not accepted in the practice of mainstream medicine and do not represent the consensus of medical opinion at this time. The reader should be aware of this and the author and publisher are not responsible for errors or omissions, or for the consequences from the application of the information in this book and make no warranty, express or implied, with respect to its contents. In particular, the author has made every effort to insure that the selection and dosage of natural substances in this book is in accordance with current recommendations. However, in view of ongoing research and the increasing understanding of the actions of natural substances and their interactions with other medications, the reader is urged to thoroughly research such choices before administration, especially when the recommended substance is an infrequently employed one, or totally new to the practitioner. It is the responsibility of each provider to ascertain the current status of, and possible complications from, any and all materials that are dispensed or recommended.

Acknowledgements

Dedicated to **Donald K. Lamb**,
who considered himself
the luckiest man in the world,
and was probably right.

Many thanks to
Karen A. Woodfork, Ph.D.,
for her invaluable assistance
in making this book possible.

Also by Dr. C.P. Negri:
Green Medicine
Naturopathic Treatment of Blood Pressure
The Negri Manual of Natural Medicine

Contents

Foreword

How many times do you go to a doctor only to hear, "Take this pill, go get this test, stop smoking, eat better and exercise more"? The physician likely spends just 15 minutes with you in an attempt to figure out what is best for you, your body and mind. While this may work for some, I have seen far too many patients who are frustrated by this process. In many cases, tests come back negative. The patient is still in distress with symptoms but told by their physician that, "It's all in your head. Go see the shrink."

More and more evidence is showing that it is *not* all in your head, and that the link between your body and your mind is very powerful. There is a need in our culture to shift the medical paradigm from treating only the symptoms to treating the whole person.

Evidence for this need is astounding. In the United States, we have the highest rates of obesity[1], diabetes and depression[2] in the world. This is not by accident. Our world has become full of "stress" at all levels: physical, emotional, mental and spiritual. We have lost our nuclear families. Electronic devices with toxic waves are all around us. Fluorescent lights are in our offices and schools[3]. We have too many cars on the road emitting noise and air pollution. The pesticides and hormones we use to grow our food change and damage our cells. We can't pronounce half of the ingredients in our food. We have toxins in our water linked to brain cancers. We don't slow down to relax. We overwork, are underpaid, and yet we are a culture of overachievers and over

---

[1] 30.6% in USA compared to 3.2% in Japan & S. Korea (2005 OECD Health Data)

[2] 9.6% of the population experiencing bipolar disorder, major depressive disorder or chronic minor depression over the course of a year compared with a 0.8% rate documented in Nigeria. (2004 WHO & Harvard Medical School study)

[3] Increased migraines, fatigue, anxiety, lower libido and possible hypertension in some populations due to increased UVB emissions and obvious lack of full spectrum lighting; See work by Dr. Ott and others.

indulgers. As a result, we have been conditioned to rely on the medical professional to do the right test and give us the right pill to make us "feel better".

Most physicians are still attempting to treat patients by addressing the symptoms alone. As an allopathically trained psychiatrist who now practices integrative, holistic medicine, I am thrilled to see how similar the naturopathic method of treatment is to how I practice. I discovered years ago that my patients were not seeing improvement in their symptoms with medications or therapy alone and that they needed more. It become apparent that there were underlying 'root' issues that needed healing before any real cures could be achieved. Getting to this root involved a process of eliminating toxins, unblocking energy, balancing the endocrine system, and generating healthy, balanced nutritional input.

This approach requires that the patient is invested in their health and treatment. This is a partnership between patient and practitioner: another factor that is evolving from the conventional, paternalistic model of medicine. It is an opportunity for the integrative, holistic or naturopathic healer to engage the client as an active participant in their treatment. This process represents fundamental pillars of both naturopathic and holistic medicine: people have an innate wisdom to heal themselves.

This book is a refreshing and much needed addition to the now growing armament of literature on integrated treatment of mental illness. I applaud Dr. Negri for this undertaking to explain and bring together not only the treatments for emotional illness, but how to integrate and think through a case. This most valuable resource will be a tool for practitioners to learn how to bring together a complicated body of knowledge and apply it to psychological cases.

Lynn M. Klimo, MD, DCEP
Integrative Psychiatrist
Co-Founder of The Center for Integrative Psychiatry and Wellness, Inc.

Contributing Author's Foreword

I was excited to hear that my friend and colleague Dr. Negri was writing about naturopathic treatment of mental and emotional disorders. Knowing him to be an assiduous researcher and an expert in many areas of alternative medicine, I expected the completed work to be eclectic, accurate, and thought provoking. With the current Western emphasis on pharmacology and widespread ignorance of most natural approaches, this work could be one more stepping stone to integration and transformation in our health care system. When he asked me to contribute to the chapters on manual therapies and body-centered psychology, I was honored.

Years ago, while earning a master's degree in counseling at Kent State University, I realized the importance of addressing the physical body when dealing with emotional issues. Talk therapy began to feel inadequate, and I was drawn to massage therapy in order to obtain a license to touch.

In the following few decades spent studying and practicing massage therapy, polarity therapy, craniosacral therapy and kinesiology, I have repeatedly listened to clients describe how some long-term emotional issue or trauma was resolved during a bodywork session, often after years of conventional psychotherapy had failed to provide relief. I have also worked in conjunction with psychiatrists and psychologists to provide adjunctive treatment for patients with various issues. The value of touch therapy in mental and emotional health is undeniable. Only the details and mechanisms are yet to be completely understood.

The body-centered methods described in this book are not meant to be exhaustive. They are included to stimulate thought, broaden awareness and provide more tools for modern clinicians. My sincere hope is that the effort will be successful.

Randall Gibson, M.Ed., LMT

Preface

Anyone who has been in family medicine for decades knows that he is sometimes a father-confessor, sometimes a friend, and *always* a psychologist, regardless of his training. People never seem to bring anything small into a doctor's office, and especially if it is a natural medicine practice. The very nature of our medical art, with its slower pace, attention to detail, and treating the person as an individual, seems to encourage the person with a problem to seek answers in our offices. It may be a strained back that brings them in the door, but it is the strain of everyday life from which they are really seeking relief.

Over the last several years, I have noticed a sharp increase in anxiety and depression among my patient population. Increased also is the number of complaints the average visitor to our clinic brings with him—it seems no one has *one* health problem anymore. Sometimes the list grows while the patient is there in the exam room! It is clear that people are unhappy, and that their emotional ills are being translated into physical ones (and vice-versa). For practitioners who did not set out to be psychotherapists, it can often be overwhelming. But, just as the auto mechanics of today must use computers, my colleagues and I must accept that we have to be able to address those mental and emotional conditions. And we must do it as effectively as we treat allergies, digestive disturbances, joint pain, or any of the typical problems people consult us for.

Fortunately, there are many effective options from our natural armamentarium. The public is ready for them; the science establishment is admitting that the antidepressants people have been taking often don't actually work. Pills that have been prescribed like candy are now known to have little real science behind them. In light of this, it is my hope that the practitioner of natural mental healthcare can now ignore cries of "quackery!" and go right ahead with ministrations of light, oils, herbs, diet, acupuncture, flower extracts, homeopathic medicines, and body-based psychotherapy. After all, statistics indicate that they will be *at least* as effective as synthetic drugs.

C.P. Negri, OMD, NMD

1

OVERVIEW of THERAPEUTICS

To fully understand the beauty of the naturopathic viewpoint, one must first reflect on the history of psychiatric treatment.

In the beginning, treatments for psychoemotional illnesses were few. These disorders were categorized—neurotic, psychotic, committable, intractable, non-fatal—and were simply managed. Protocols were developed over time that had some impact on these states by changing something within the physiology, and secondarily, the patient's behavior. These became the long-used treatments in mental health facilities that are now archaic: strapping the person down for long periods, mild electric shocks, extended baths, cold sprays and sheet packs, etc. The amount of time and effort required, coupled with the minimal efficacy, led the profession to look for more direct ways to impact disordered mind states than beating and chaining down those who possessed them.

This resulted in more aggressive methods being adopted. Now the uterus would be removed. Strong electric shocks were applied to the brain to produce convulsions. Whole sections of brain tissue were gouged out. People weren't healthier, but the offending behavior was often eliminated. Those who were not sufficiently managed in these ways were still locked away at the behest of their family members.

Then, purely somatic therapies gave way to psychopharmacology. The known effects of opiate drugs (and of course, alcohol) on emotions, thought processes and behavior were well documented. It remained to find new types of oral agents by

1

which psychological problems could be managed. The brain operations, chaining, shocking, and dunking could then be abandoned.

Pharmacology from this point on played an increasing role in mental health methods. It came into its own after World War II and advanced steadily into the 1960s. This approach was seen as more humane, if not more effective; however, treatments like electroconvulsive shock therapy continued to be used. Psychoanalysis and other forms of talk psychotherapy were available, though not utilized by the average person.

But non-drug therapy was gaining momentum and evolving. Each decade has had its unique flavor and slant toward treating psychoemotional problems: psychoanalysis in the 1940s and 1950s, behavioral therapy in the 1960s, humanistic therapies in the 1970s, cognitive and psychodynamic approaches in the 1980s, and the emergence of "energy psychotherapy" methods in the 1990s. These variations reflect changes in culture and society as much as, or perhaps more than, changes in the mental health professions.

The tide changed for the treatment of psychoemotional illness, not by advances in medical methods, but by changes in popular culture. Up until the 1960s, people in our society did not typically discuss their personal problems openly. The '60s changed all that. Because of the explosion of self-help therapies that incubated during that decade, and the general movement toward openness in society, things altered dramatically. The exploration of various (typically Eastern) approaches to spirituality became popular, and the role of psychology moved from the Freudian analytic model to the far more accessible and practical forms of emotional healing. No longer relegated to the wood-paneled offices of intellectual psychiatrists charging high fees, therapy became something that was actually "trendy" to be pursuing. A notable example of this existed in the "encounter groups" of that period, where strangers gathered for a weekend and gradually learned to trust each other and bare their souls in an effort to work out their problems (this approach was lampooned in the 1971 hit movie, *Bob and Carol and Ted and Alice*).

Emotional problems began to lose their stigma of being equated with "craziness". Over the next couple decades, it became much more accepted for someone to be "in therapy", regardless of what that therapy was. A goodly number of new approaches to counseling and guidance developed, as well as a legion of best-selling "pop psychology" books on improving your life. This was a small genre in the publishing field until the 1970s, but it has since claimed a huge showing in the arena of printed literature.

People came to find friendship and also romantic relationships by sharing their emotional trauma. This is an important point that is given attention at various junctures in this book.

Nevertheless, the pharmaceutical management of emotional illness, particularly depression, came to be more emphasized over time than non-drug therapies in the medical community. In fact, antidepressants became one of the most widely selling drug categories.

The naturopathic armamentarium contains many psychoactive materials of great value. To use them effectively requires at least a passing knowledge of what has been done with synthetic drugs in the field. This is both for the purpose of better understanding the patient who has been taking allopathic drugs and for a better appreciation of the differing philosophy in prescribing natural medicines.

Psychopharmacology

Psychopharmacology as a distinct field became established in the early 1950s. Chlorpromazine, an antipsychotic drug developed in 1950, became the first widely prescribed psychiatric drug. As a type of synthetic antihistamine, it was noticed to have psychiatric effects, and its changed usage from allergy drug to antipsychotic drug generated much income for its French manufacturer. Its lucrative success led to derivative drugs by other companies.

Because anxiety was a key component of many psychiatric complaints, tranquilizing drugs began to take prominence in treatment. Although tranquilizing medicines had been around for centuries in the form of botanicals, they were tranquilizers only in the sense that they usually put people to sleep; they were, technically, *soporifics*. That changed when a new class of drug was invented in 1954. The meprobamates were unique in that they calmed without sedating. The term "tranquilizer" became part of the lexicon, and the two stars of this world were Miltown and Equanil. You never hear of these drugs today, but they are representatives of a category originally known as "minor tranquilizers" (as opposed to "major" tranquilizers like Thorazine, prescribed for severe emotional illness and psychosis). Minor tranquilizers are referred to today, perhaps more accurately, as *anxiolytic* (anti-anxiety) agents.

Meprobamates became overshadowed by a new class of minor tranquilizer/anti-anxiety drugs, the benzodiazepines. The first, Librium, was discovered accidentally in 1955. Benzodiazepines are not only anxiolytic; they are sedative, hypnotic, anticonvulsive, and skeletal muscle relaxants. Dependency is high with this class of drug. Diazepam (Valium) is one of the most widely prescribed drugs in history—and one of the most common addictions. Lorazepam (Ativan) is used often for panic attacks. Some forms of benzodiazepine drugs, such as alprazolam (Xanax), are used for anxiety that is coupled with moderate depression. Which brings us to the next category.

Depression began to be more commonplace in the 1950s. Drugs to act specifically on this problem were developed. Monoamine oxidase (MAO) inhibitor drugs became available in the late '50s. These antidepressant drugs were said to prevent monoamine oxidase, an enzyme, from breaking down neurotransmitters in the brain; when these brain chemicals were in greater supply, the theory went, the mood was higher.

4

Tricyclic antidepressants appeared in the same time period. These drugs block the transporter for norepinephrine and serotonin, thereby preventing their re-uptake and insuring that greater amounts of these neurotransmitters remain in the brain. Again, the expected result is an elevated mood. Tricyclics were thought to be more effective than MAO inhibitors and have fewer side effects, so prescriptions of this class of drug increased.

These classes of drugs were not the first to be used for psychoemotional problems, however. The barbiturate-derivative drug Secobarbital had been actually been around since the 1920s but came to be popularly prescribed in the 1960s and 1970s due to its sedative and hypnotic effects. It had an impact on anxiety, and it masked poor coping patterns by creating a drugged, hazy outlook and slowed thought processes. Widely used as a street drug as well, it came into general familiarity by being featured prominently in a best-selling novel, *Valley of the Dolls*. It was responsible for the overdose deaths of entertainer Judy Garland and musician Jimi Hendrix, among many others. Between 1964 and 1970, prescriptions for psychotherapeutic drugs increased by a steady seven per cent a year, and more than a third of these were anti-anxiety drugs[4].

Then came the selective serotonin reuptake inhibitor (SSRI) drugs, developed in the 1990s. These specifically prevent only the neurotransmitter serotonin from being taken up into the cells, once again increasing the amount of the brain chemical available to bind to its receptor. SSRIs became widely prescribed for depression and anxiety disorders[5]. In fact, they are the most popularly dispensed psychoactive drugs in many countries. Tricyclics and MAO inhibitors came to be less often prescribed, although they are still preferred in certain cases by many clinicians.

---

[4] Balter, Mitchell, and Levine, Jerome, "Character and Extent of Psychotherapeutic Drug Usage in the United States," presented at the Fifth World Congress on Psychiatry, Mexico City, Nov. 30, 1971

[5] Also eating disorders, obsessive-compulsive disorder (OCD) and post-traumatic stress disorder (PTSD).

In the first ten years after SSRI drugs debuted, prescriptions for depression increased 73 percent in adults and 100 percent in the elderly. Prescriptions for *children* increased by fifty percent. By 2005, almost ten percent of Americans of all ages (more than 27 million people) were taking antidepressants. Altogether, sales for antidepressant drugs in the U.S. totaled more than $10 billion by the year 2008.

Besides school-age children, populations that pharmaceutical companies have targeted for these drugs are military personnel and pregnant women (23% of whom become depressed, statistically). **This is despite the known effect of SSRI drugs increasing the risk of suicidal behavior by five times**[6]. Nevertheless, the American Psychiatric Association and the American College of Obstetricians and Gynecologists have jointly endorsed the medicating of pregnant women. And while untreated depression (or ineffectual therapy for depression) can increase complications of pregnancy, antidepressant drugs carry the increased risk of fetal deformities, deadly pulmonary hypertension, respiratory difficulty, and all the withdrawal symptoms that the mother exhibits when discontinuing the drug.

Depression in pregnant women, though, is pointing to factors in that disorder that are not limited to that population group. Nonpregnant women, and men also, fall into depression for the same reason that women do when they are expecting a child. That reason is hormonal imbalance. The implications of this will be discussed shortly.

While many more rational authorities in the psychiatric field suggest that psychotherapy in the form of counseling and cognitive therapies be tried before drugs, it matters little. This plan of action is not often pursued, for two main reasons.

Once again, people associate psychotherapy with the stigma of mental illness and are less likely to indulge in it than swallowing

---

[6] *J Am Phys Surg* 09; 14:7-12

a pill. After all, everyone they know is doing that already for other health problems; the culture supports it. But the other limiting factor is that insurance plans most often do not cover psychotherapy services. They will however, cover the drugs, which would otherwise be unaffordable. The psychology of getting something expensive for less money is well understood by the psychiatric profession and the pharmaceutical industry. Therefore, we have a situation where last resort measures are tried first, and (as elsewhere in medicine) the most gentle, natural, and at times most effective therapies are tried last.

On the other hand, the various approaches to non-drug psychotherapy have established themselves as a collective cultural institution. Not only has a large percentage of the American public had at least a fleeting direct experience with some type of therapy, the *concept* of therapy—and the pathologies it is intended to help—is now part of the national consciousness.

Drugs vs. Talk

There has existed for some time a rift between psychopharmacological approaches to therapy, exemplified by psychiatrists, and the various analysis and counseling methods, as embodied by psychologists and counselors. Proponents of the former have placed such emphasis on brain studies and the biochemical model that they seemingly reduce talk therapy to an adjunctive role; the latter camp cites studies that indicate that their therapy is not only effective, it is statistically *more effective* than drugs in balancing psychoemotional illness.

Indeed, the most up to date meta-analyses completely refute the efficacy of the most often-prescribed drugs (SSRIs), revealing that **their actions statistically have no more effect than placebo.** While there appears to be pharmacological action on the most severe depression, the effects for mild and moderate depression are "minimal or non-existent."[7] Years before SSRI drugs became commonly prescribed, a study[8] found the then-popular tricyclic antidepressant Imipramine ineffective in much the same way. In sixteen weeks of randomly assigned treatment, lasting improvement was gained by 30% of those treated with cognitive therapy, 26% of those treated with interpersonal therapy, and only 19% of those treated with Imipramine. To put this in perspective, 20 % of those in the placebo group improved, compared with 19% in the drug group. **The antidepressant was outperformed by a placebo.** What is astounding is that the authors of the study summarized the trials as showing that the results "did not differ significantly among the four treatments." Does someone really think that there is no significant difference between thirty percent and nineteen percent?

Talk therapy now appears to be a much more valid approach in most cases.

---

[7] The following studies showing little support for antidepressants were published in peer-review journals:
- Fournier J., DeRubeis, R., Hollon, S., et al.; Antidepressant Drug Effects and Depression Severity, A Patient-Level Meta-Analysis, *JAMA.* 2010; 303 (1): 47-53.
- Kirsch, I., Sapirstein, G.; Listening to Prozac But Hearing Placebo: A Meta-Analysis of Antidepressant Medication, *Prevention & Treatment* (1998) 1: Article 0002a. http://journals.apa.org/prevention/volume 1/pre0010002a.html.
- Hollon S., DeRubeis R., Shelton R., Weiss B.; The Emperor's New Drugs: Effect Size and Moderation Effects. *Prevention & Treatment* (2002) 5: Article 28.
- Kirsch I., Scoboria A., Moore T.; Antidepressants and Placebos: Secrets, Revelations, and Unanswered Questions. *Prevention & Treatment* (2002b) 5: Article 33.

[8] Shea MT, Elkin I, Imber SD, Sotsky SM, Watkins JT, Collins JF, Pilkonis PA, Beckham E, Glass DR, Dolan RT, et al., Course of depressive symptoms over follow-up. Findings from the National Institute of Mental Health Treatment of Depression Collaborative Research Program, *Arch Gen Psychiatry.* 1992 Oct; 49 (10):782-7.

Yet, of the multitude of approaches to psychotherapy[9], study after study has failed to demonstrate the superiority of any particular school of thought. Cognitive therapy appears to have an advantage in treating distortions of the thought processes, while behavioral therapy seems to be especially efficacious in treating specific symptom complexes. Phobias, anxiety attacks, and obsessive-compulsive disorders seem to respond well to a combination. A 1989 National Institute of Mental Health study showed interpersonal therapy slightly edging out cognitive therapy in efficacy in the treatment of depression, and both competing favorably with antidepressant drugs.

Most therapists today do not cling to a rigid dogma of a particular school of thought, but rather they embrace an eclectic mix of technics that were born of various theories. This "integrative" approach allows a more adaptive handling of the client's problems, and makes possible a more individualized therapy and therefore (in all likelihood) increases the overall success rate of a particular therapist's practice beyond what it would have been in the days of a strict adherence to Jungian, Adlerian, etc., methodology.

The Contemporary Standing of Therapy

Freud would never have believed the extent to which the seeds he planted here in the early 1900s have taken root. We now have a society that speaks the language of psychotherapy, from the casual use of clinical terms like "passive-aggressive" by the lay public to the vague and seemingly common sense platitudes "getting in touch with your feelings" and "talking to your inner child". Difficult people are said to "have issues". Self-help sections of bookstores have grown beyond belief. M. Scott Peck's book *The Road Less Traveled* remained on the *New York Times* bestseller list for ten years, an unprecedented achievement. Television saw the discharging of emotional distress as public entertainment with the advent of the Oprah Winfrey show, and many have followed since then.

---

[9] Estimated to number over four hundred, according to Dr. T. Byram Karasu, psychiatrist at Albert Einstein College of Medicine.

While many opt for psychotherapy to improve their lives (and probably everyone would benefit from some amount of therapy), a far greater number who are in need do not seek help. It is possible that the very kinship they feel with those airing their troubles on television creates a type of "virtual group therapy". It is also possible that the incessant publicizing of celebrities and their emotional difficulties, addictions, maladaptive behavior, etc., has resulted in a lowering of standards for social interaction and personal happiness. After all, if the rich and famous with every advantage available to them are just as miserable as us, why bother?

But how does natural medicine interface with this schism between the biological and the sociological? With psychotherapists applying tried and true methods of non-drug treatment, and psychiatrists routinely applying pharmaceuticals to reduce the whole matter to an alignment of molecules in the brain, what possible role do naturopaths play? What place do their remedies occupy in this already controversial field? The average person may see some possibilities in herbal medicines and their chemical effects, but physical therapies using light, sound, touch, water, electricity, and other modalities may seem ridiculously misapplied to the realm of mental and emotional disorders. The simple fact is that they are not.

2

THE NATUROPATHIC VIEWPOINT

One must begin with the basic tenets of naturopathic medicine:

I.   Utilize the inherent ability of the person to heal
II. Identify and treat the cause, not result, of disease
III. Do no harm
IV. Treat the whole person
V.   Teach health maintenance and encourage responsibility for it
VI. Seek to prevent, rather than treat, further problems

The traditional naturopathic perspective of illness, reaching back to the early 20th Century, is that disease is caused by a buildup of toxins in the body from improper living. Treatment for nearly any illness therefore involves clearing such toxins from the system and rebuilding tissue and function with the elements needed to accomplish that.

Now to reduce the complexity of pathology to a simplistic concept of "toxicity"—as though the vaguely identified "toxins" could result in hundreds of completely disparate diseases— appears absurd in light of modern medical science. But before this concept is discarded outright by the reader, it must be pointed out that:

- Excessive levels of, or poorly processed, hormones become toxic to the body. When clearance pathways are not efficient, the hormones have a negative impact on physical as well as mental health. They become, in effect, toxins.

- Phthalates, used in making plastics, are known to be responsible for a multitude of health problems, most notably those involving hormone imbalances. Although the United States instituted a ban on phthalate-containing childcare articles and children's toys in February 2009, many personal care products still contain undesirable levels of these chemicals. The ability of these substances to mimic estrogens and androgens has led to negative changes in brain neurotransmitter and hormone precursor levels, making them capable of causing emotional and cognitive illness as well as physical illness. To call these "toxins" is not unreasonable. To have these substances leaching into our food and water from plastic containers *is* unreasonable.

- Chemical and metallic residues are encountered daily from contaminated air and drinking water. Given what we know about their effects, public health experts refer to these, accurately, as "toxins".

- The outgassing from plastics, synthetic fibers in carpet, formaldehyde solutes from dry-cleaned clothing, and many other sources in the home add to the toxic soup that is modern-day life.

- A car dashboard is known to emit as many as 200 volatile organic compounds (VOCs), including C3- and C4-alkylbenzenes, and aldehydes. Car exhaust can contain the known carcinogens 1,3-butadiene, benzene, and formaldehyde. Sealing oneself in this toxic capsule repeatedly, if it does not result in measurable changes in blood composition or outright illness, certainly can contribute to more subjective changes in neurotransmitter function and therefore mood, etc. Remember that it takes far less exposure to a given trigger to provoke a change in mood than a change in one's blood count.

- Microbes tend to migrate to areas that are already toxic. Infections, even subclinical ones, necessarily involve the further deposition of toxins in the tissues by the invading organisms. This is true not only of living organisms, but killed ones as well. While the active infection is considered gone after antimicrobial treatment (because the bacterial or viral count is down), one's body is still dealing with the aftermath. The breakdown of killed microbial bodies requires energy from the body as well as a working clearance mechanism.

- Reactions from food sensitivities create cascades of chemicals that can be toxic to the tissues, as part of the body's response to what it interprets as an assault from without. Changes in neurotransmitter function in the brain are known to occur with not only food additives, but also perfectly healthy foods that the person happens to be allergic to.

- The known effects of low-frequency electromagnetic pollution in the home include decreased pineal gland function (with resultant decreased melatonin production), cognitive thinking disorders, and an increase in leukemia and cancer rates. Appliances (hair dryers, shavers), fluorescent lights, computers, cell phones, and current-drawing lamps, alarm clocks, and TV sets close to the bed add to any other exposure to EMG fields elsewhere. The impact on mental/emotional health is considerable.

- The research of Dietrich Klinghardt, M.D.[10], suggests that specific toxins in the body will couple with specific psychoemotional problems. The emerging science of neurobiology is advancing steadily, suggesting among other things that information regarding a particular issue in the patient's mind can embed itself in a toxin that is harbored by the tissues. Clearing the issue, then, necessarily involves clearing that toxin from the body.

All the factors just mentioned can affect psychoemotional states.

---

[10] Klinghardt Academy of Neurobiology, LLC,
4630 Talbot Drive, Boulder, CO 80303, (303) 499-4700

Phase I reaction (composed primarily of the cytochrome P450 supergene family of enzymes) and Phase II conjugation in the liver are considered part of the body's highly complex detoxification system. Since the liver is one of the most overburdened organs in modern life, due to the massive number of *xenobiotics* (external-source agents) we encounter, one must conclude that without aid, this system of toxin clearance may not be able to do its job properly. Therefore, the traditional naturopathic concept of toxicity is more valid than it appears at first glance. The use of antiquated language and simple principles of operation in no way reduces the legitimacy and efficacy of natural medicine procedures.

Naturopathic Modalities

Naturopathic medicine has historically contained an eclectic array of treatment methods. After naturopaths refined their protocols for treating illness with diet, exercise, light, and applications of water in the late 1800s, manual therapies such as massage and joint manipulation were added in the early years of the 20th Century; nutritional supplements in the 1920s; electrotherapies and botanical medicines in the 1930s; homeopathic medicines in the 1940s and 1950s; holistic counseling in the 1970s; and acupuncture in the 1980s. Natural medicine has evolved from a number of unrelated empirical methods to a cohesive school with an underlying philosophy that unites them.

All these modalities can have an impact on psychoemotional illness and will be covered in succeeding chapters. The patient consulting a practitioner of natural medicine who can orchestrate these various methods stands a greater chance of lasting benefits than those treated with one method, not to mention the nearly total absence of side effects.

In forthcoming chapters, the following modalities are detailed, centering on their application to psychoemotional illnesses:

- Nutrition: Diet and Orthomolecular Therapy
- Phototherapies
- Acupuncture
- Manual Therapies
- Body-centered Psychology
- Olfactory Therapy
- Flower Essences
- Biological Medicines
- Botanical medicines
- Homeopathic Medicines
- Integrated Methodology of the above

But before examining these, an important point must be made. The paradox exists, as we now know, that while serotonin reuptake inhibitors seem to benefit some people with depression, other drugs that have minimal action on serotonin (such as Desipiramine, a tricyclic) also appear to help depression. If serotonin plays an essential role in the etiology of depression in one case, how can it be irrelevant in the other case? The world of orthodox medicine can wonder, but we in the natural medicine field can understand from history just why this is so.

When naturopaths a hundred years ago treated microbes as though they did not exist as they detoxified their patients, they assumed the germ theory was in error because their patients got better. Science had not yet discovered the complex factors of immunological response that enabled those patients to fight infections—a response that was set in motion by the doctor's ministrations. Not only somatic disease but also deeply seated mental disorders yielded to such simple procedures as fasting[11].

When the early chiropractors touted the vertebral subluxation as the cause of diseases, they were holding to a theory that would

---

[11] As recently as the 1960s, there was a well-documented study of the efficacy of fasting (Nicolayev, Y.S., *Controlled Fasting Cure of Schizophrenia*, Moscow 1963)

15

not stand the test of time and science. Yet, organic disease cleared time and time again in patients who had their spines manipulated. The myriad of chemical and subtle changes that the seemingly mechanical chiropractic adjustment caused was enough to nudge the patients' systems in a healing direction.

Was it the placebo response at work in these two examples? If we say yes to this question, it implies that the patients only *thought* they got better. The fact is that objective testing has corroborated many thousands of cures at the hands of unorthodox healers who used methods that should not be effective in light of contemporary knowledge. But if we say no, it is not the placebo response at all—this is a denial of a most powerful force that is activated by *every* health care practitioner, orthodox or unorthodox.

It is likely that what we today call the placebo response is a part of a more complex mechanism that is the core of healing itself. The simple methods of the old-time naturopath and chiropractor as just described caused a mild shock to the system under controlled conditions—conditions that included intent to heal on the part of the doctor and intent to *be* healed on the part of the patients. It is likely that neither intent alone nor action alone is entirely satisfactory. Applying a force that stimulates the body-mind system into changing its current course is probably the essence of therapy. Jogging the healing response in non-specific ways is clearly effective. This is why people continue to get well from having their spines manipulated, from fasting, from being sprayed with cold water, from having needles inserted into their limbs (sometimes even in the wrong spots!), and from taking medicines seemingly too weak to have any action.

That even non-physical forces such as psychotherapy can accomplish this is recognized by Jerome D. Frank, MD, in his book, *Persuasion and Healing: A Comparative Study of Psychology*:[12]

---

[12] 1961 Johns Hopkins Press, Baltimore

Certain types of therapy rely primarily on the healer's ability to mobilize healing forces in the sufferer by psychological means. These forms of treatment may be generically termed psychotherapy.

Consider that:

- A sudden emotional shock has eliminated long-standing illnesses in many people. It has also spontaneously *caused* diseases in other people.

- It is often observed that chronic illnesses clear up for the duration of a woman's pregnancy. It is not possible to explain this purely in terms of the hormonal changes of gestation.

- There are cultural rituals in third world countries that are used for those who experience emotional trauma. Villagers who have suffered mentally, or have physical symptoms stemming from an emotional event, are treated by a cleansing ceremony in which they are deliberately terrified. The healers pose as evil spirits that, according to legend, visit the sick. The "patient" is made to scream, dance, and shout, often speaking in tongues. The cleansing can last more than twenty-four hours. By frightening the sick person in this way, spectacular recoveries are often seen. In its way, this appears to utilize the principles of Homeopathy (see Chapter Thirteen) by curing like with like. But the idea of a counter-shock has many different applications.

3

PATHOLOGY and ASSESSMENT

Using names of disorders to pigeonhole people is inaccurate and ineffective. As with the classical homeopathic perspective (and holistic medicine in general), the "pathology"—if any exists at all—is not what is to be treated. The person who is having trouble coping or modifying his or her behavior in a positive way should be the target of the therapy, not the disease entity that is supposedly causing the problem. And, like somatic illnesses, the shortest route to finding the effective therapy is to examine exactly how the problem is manifesting in the person: the individualizing symptoms. Which symptoms are revealing of that particular person's personal "syndrome"? While conventional medicine arrives at a diagnosis when the common symptoms of the illness appear in the patient, we must pass by the symptoms or characteristics that are common to the diagnosis and focus instead on those that are individualizing. In one particular person, a symptom or set of symptoms may be caused by something—and relieved by something—completely different from the next patient with the same diagnosis or symptoms.

John R. Lion, Clinical Professor of Psychiatry at the University of Maryland, has said, "Psychiatry has tried so hard to fashion itself as a medical discipline that it has shot itself in the foot. Recovery from mental illness does not obey the laws of physical illness."[13]

---

[13] "Please Make the Mental Patient Go Away"; *The New Psychiatric Review* (Sheppard Pratt Health System), 3: 3 (April) 2001

The Psychiatric Viewpoint

From the psychiatric viewpoint, diagnosis and treatment of mood disorders are two side of the same coin. Diagnosing a mental or emotional illness demands an assembled list of characteristic symptoms.

The most recognized and utilized collection of symptoms and criteria of diagnosis is found in the Diagnostic and Statistical Manual of Mental Disorders (DSM), published by the American Psychiatric Association.

The generic answer to the question "What is anxiety?" or "What is depression?" or "What is bipolar affective disorder?" is a uniform answer: It is an illness, diagnosed by the presence of certain symptoms, and treated with the appropriate drugs. Treatment is first and foremost with oral medication, although there are other treatments endorsed by the APA.

This viewpoint is highly flawed for several reasons:

1. The lack of uniformity in individual symptoms (different people with the same diagnosis may have vastly different manifestations)

2. The lack of a solid etiology for any given disorder; what may cause a condition in one person causes something else in another person

3. Even if the drugs used were as effective as their claims—which have now been found to *not* be the case—they would still be simply antidotal to symptoms, and not curative.

Psychiatry has bolstered its confidence and image by claiming science as its ally. It is now explaining all phenomena as the effects of "brain chemistry". By exploring which areas in the brain are involved with particular emotions and thoughts, by isolating specific neurotransmitters, by mapping out genes that carry programming for characteristics—all the myriad manifestations of the human mind can be put down to "brain chemistry".

Depression is an imbalance of a chemical in the brain. Love is the presence of a chemical in the brain. Spiritual feelings are an activation of a region in the brain.

While these findings (and any findings that increase understanding of the human condition) are welcome, they can easily serve to distract us from the real issue at hand, and the increased knowledge can actually leave us less empowered than before. To interface with these unwanted changes in the brain, we are dependent on pharmaceutical companies to provide the tools to straighten out whatever imbalance that we find with our brain imaging and lab tests. With the claim "It's all chemistry" we are made to feel that the solution is totally chemical. While this may be true at the end stage (i.e., positive changes in the brain will be accompanied by changes in brain chemistry) it is not necessarily true of the methodology. Non-drug interventions change brain chemicals, too.

First-tier Assessment

Assessment of the patient's mental and emotional status begins with good observation. Apart from any information disclosed by the patient, the examiner must watch for any signs that indicate a particular deviation from the "norm". *This is not for the purpose of establishing "pathology"*, but rather for finding those characteristics that are so distinctive about the individual that a precise approach can be taken in restoring the person to full health.

The following checklist is a standard one to start with, in order to arrive at an initial assessment of the patient with psychoemotional problems, and also for documenting purposes. While these descriptions are satisfactory from the standpoint of orthodox charting, they serve us with only the most superficial description of the person. Nonetheless, it is possible to use the checklist to hone in on the more important information that will lead to a precise therapy. One can save time by using the quick sketch to guide the interview process.

Appearance	• Appropriate • Inappropriate • Clean and well groomed • Disheveled • Bizarre • Physically impaired
Motor Activity	• Gait normal • Gait abnormal • Tic present • Tremors present • Gestures, mannerisms exaggerated • Echopraxia • Agitation • Restlessness • Hyperactivity • Aggressiveness • Rigidity
Speech	• Within normal limits • Loud • Soft • Delayed • Excessive • Rapid • Slow • Slurred • Aphasia • Dysphasia • Stuttering • Impairment
General Attitude	• Pleasant and cooperative • Cooperative • Uncooperative • Pleasant • Hostile • Defensive • Guarded/suspicious • Apathetic

Mood	• Consistent • Neutral • Labile • Irritable • Anxious • Fearful • Despairing • Sad • Depressed • Elated • Euphoric
Affect	• Consistent with mood • Inappropriate for mood • Neutral • Sad • Flat • Empty, withdrawn • Tearful • Elated • Blunt • Hostile • Appropriate to thought • Inappropriate to thought
Thought Processes	• Attention span good • Attention span poor • Blocking • Tangentiality/loose associations • Neologisms • Word salads • Echolalia
Thought Contents	• Obsessive • Phobic • Delusional • Suspicious • Paranoid • Homicidal • Suicidal • Religiosity • Magical

Perceptual Disturbances	• Hallucinations: • Auditory • Visual • Tactile • Olfactory • Gustatory • Illusions: Depersonalization • Derealization
Sensory / Cognitive	• Alertness good • Alertness poor • Orientation x 3 • Disoriented as to time • Disoriented as to person • Disoriented as to place • Memory adequate • Memory poor • Abstract reasoning good • Abstract reasoning poor
Impulse Control	• Impulse control adequate • Uncontrolled affection • Uncontrolled sexual urges • Uncontrolled aggression • Uncontrolled fear • Uncontrolled guilt • Uncontrolled substance abuse
Judgment / Insight	• Decision-making ability good • Decision-making ability poor • Problem-solving ability good • Problem-solving ability poor • Coping skills good • Coping skills poor

Second-tier Assessment
Useful Questions to Delineate the Case

- What emotions do you have the most trouble dealing with?
- Do you have any frustrated ambitions?
- What characteristics prevent you from being exactly the person you would like to be?
- What creates the greatest stress in your life?
- How would a friend describe you?
- How would someone who does not like you describe you?
- What thing makes you the angriest?
- What thing makes you the saddest?
- What is the scariest thing in the world to you?
- What do you find yourself thinking about more than anything else?
- What do you worry about the most?
- What makes you the happiest?
- What kind of things do you do rapidly?
- What kind of things do you do slowly?

In the orthodox mental health field, the Minnesota Multiphasic Personality Inventory (MMPI) Test has been the standard method of assessing personality disorders. A psychological test consisting of 500 questions that the subject places into three categories as they apply to him, it is designed to give the mental health professional an accurate picture of the subject's personality in one session. It is useful for better defining and categorizing a patient's mental status, but lends little valuable information that can lead to a plan for true healing.

Since the orthodox medical field defines health as the absence of symptoms, it follows that medicinal approaches to mental health have focused on symptomatic treatment. While many effective forms of psychotherapy have been developed, they require time and effort, and the allopathic practitioner is pressured to produce quick results; drug therapy is commonly resorted to.

The "alternative" practitioner, as well as orthodox practitioners, will see patients with a mix of somatic and psychoemotional complaints. Unlike the orthodox practitioner, however, a line is not drawn between the two. Mental problems are not treated independently of physical problems because they are *not* independent. For this reason, the author (CPN) has always preferred the term *somato-psychic* to the commonly used *psychosomatic*, because the latter term has been mistakenly confused by the public to describe an imaginary illness. In reality, nearly *all* disease is somato-psychic to some degree, because there is almost nothing physical that does not impact on us mentally, and vice-versa.

Wilhelm Reich, M.D., a pioneering psychotherapist who took a global view of the patient, originally called his method *vegeto-therapy*. It was based on the premise that emotions are not only cortically manifested, but actually emanate from organic and musculoskeletal systems. Emotional tension is accompanied by neuromuscular tension. Aspects of the personality and behavior are functionally identical to bodily attitudes. The effective psychotherapist, therefore, must analyze and re-train the neuromuscular patterns as well as the emotional ones. Many systems of body-based psychotherapy, whether they consider themselves descended from Reich's work or not, have this theory at their core. These are discussed in Chapter Eight.

Detoxification and the Law of Healing

A core concept in natural medicine is that of detoxification. Methods used can be general, such as fasting or hyperthermia (from steam baths, etc.), or specific, as in using homeopathic microdoses of causative agents (i.e., chemical toxins or heavy metals) to target those substances for excretion[14].

Detoxification methods typically release toxic materials from the tissues, which the body then expels through the pores, the bowels, in the urine and even the breath (a separate process further

---

[14] Animal studies have shown that microdoses of arsenic and lead have stimulated greater excretion of those toxic substances in the urine than in control groups.

defined as *drainage*). Symptoms often arise during this process, classically known as the *healing crisis*. The drainage of waste material, coming to the surface, can create symptoms that resemble the symptoms of acute illness. This is often misunderstood by the patient as a new problem when he is actually being rid of old disease byproducts.

There has an observation throughout the history of natural medicine that the healing response creates a re-awakening of old disease patterns in the individual, and that in the process of the healing crisis, former problems are revisited. While naturopaths speak of the "healing crisis", and chiropractors describe it as "retracing", it was the homeopaths who gave it the name by which it became codified. *Hering's Law* is the observation that healing takes place in three predictable directions:

- From deep inside the body outward toward the surface;
- From the top of the body downwards;
- In reverse order of appearance.

It is the latter aspect, reverse order, which creates the classical revisiting of old illnesses in the person who is healing. Another term for this, as suggested by Dr. Hans-Heinrich Reckeweg, is *regressive vicariation*. Reckeweg is the father of the method known as Homotoxicology, a system that uses homeopathic medicines in line with orthodox biomedical concepts. It is being mentioned here because it applies to mental/emotional disease as appropriately as physical disease.

The framework in Homotoxicology describes the typical progression of psychopathy:

- Phase One: **Alteration** in the form of nervous tension and emotional reactions to circumstantial stresses.
- Phase Two: **Reaction** in the form of exogenous depression, or attention-deficit disorder.
- Phase Three: **Fixation**, a condition where phobias or neurotic depression develop.
- Phase Four: **Chronicity**, a deeper stage where endogenous depression, anxiety neurosis, and psychosis form.

- Phase Five: **Deficits** in the form of mental deficiencies and schizophrenia.
- Phase Six: **Decoupling**, in the form of mania or catatonia, the deepest level of disturbance to the mind.

Successful resolution of psycho-emotional complaints, like somatic complaints, depends on utilizing Hering's Law. Successful therapy will elicit a re-tracing backward of symptoms and issues as the individual resolves those issues with the aid of therapies that will remove obstacles to the healing.

Obstacles to mental and emotional healing, oddly enough, may be the same as those that oppose full healing of somatic complaints. It appears that every unresolved psycho-emotional conflict has an attached physical toxin that must be shed for complete clearing of the problem. Therefore, *somatic treatment is necessary for psychotherapy to be successful*. This is a statement that will be challenged by some as absurd, while others will view it as so self-evident as to be unnecessary to mention. This subject is dealt with in detail in Chapter Fourteen, *Integrated Methodology*.

Third-tier Assessment

After doing the basic psychological evaluation and asking questions to arrive at a more individualized personality/emotional profile, make note of where the person is with regard to the six phases described previously. This should be fairly simple from the history and presentation.

Energetic Anatomy

In some Western religious/metaphysical systems such as Theosophy, but far better known through the Eastern practice of Yoga, there is much discussion of the "etheric double" of the physical body. Existing in the same space, this intangible counterpart is composed of subtle matter. Within this ethereal duplicate lie the *chakras*. These are usually defined as "energy centers". A Sanskrit word meaning "wheel" or "turning", a chakra is said to be a vortex of energy rotating in a specific place in the body/bodies. They are usually depicted as wheels with spokes or flowers with petals, indicating that they radiate as well as store the life force.

This idea, of course, parts with established science. If the reader will be indulgent, it will be pointed out that the chakras are traditionally located at what we now know to be major branchings of the central nervous system. It may be that a chakra is a nexus of biophysical energy that is as measurable as other forces in the human body, but the important thing for the present discussion is that there are a number of associations with each individual chakra that have applications in emotional disorders. Since the locations of these "energy centers" coincide with known nerve plexi, the issue of whether an "etheric body" exists is irrelevant. If adherents of these metaphysical systems have for centuries observed correlations of certain mental and emotional symptoms with afflictions of particular charkas, this is something we should investigate. After all, thoughts and emotions are just as ethereal, and we have an entire industry of diagnosis and treatment based on them.

In the upcoming chapter on acupuncture, we will see how modern science has been able to confirm effects of a traditional healthcare practice that has fanciful theories that seem to run contrary to contemporary knowledge. The same may be found

29

of the chakras in the future.

In any case, the purpose of this discussion is not to argue for the truth of the chakra hypothesis. It is simply to illustrate the theories behind some phenomena that have been clinically observed. Whether the theories are ultimately proven true or not, certain treatment methodologies can be very useful in balancing psychoemotional states. Those treatments can be chosen on the basis of their correlates with known afflictions of the chakras or energy centers.

In Chapter Four, we will examine the use of light in treatment, and in particular *chromotherapy*—the use of the visible light spectrum in very specific ways. Intelligent applications of colored light have been found consistently effective in treating a wide variety of disorders, both physical and mental in nature. When a problem is originating in or affecting what could be called the "subtle anatomy", treatments such as acupuncture or light are very powerful modalities.

#	Center	Location	Gland	Influences	Functions	Underactive	Overactive	Associated Color
7	Crown	Top of cranium	Pineal	Emotions; Aspirations; Creativity; Right brain; Right side of head; Right eye	Knowledge, wisdom	Unconcerned with spiritual matters; rigid thinking	Intellectualizing; overly concerned with spirituality and neglecting of physical realities.	Violet
6	Brow	Between eyes	Pituitary	Logic; Left brain; Left eye; Ears; nose/sinuses	Vision in both the physical and the figurative senses; insight	Reliance on others or authorities rather than one's own ideas; reliance on belief systems; rigid thinking.	Illusions/delusions; living in a fantasy world.	Indigo
5	Throat	Throat	Thyroid and Parathyroid	Voice; Lungs and bronchi; Esophagus	Self-expression and also ability to hear others' expressed feelings. Lying can cause dysfunction.	Introverted, shy.	Keep people at a distance; not listening to others; domineering	Blue
4	Heart	Sternum/ between scapulae	Thymus	Heart; Vagus nerve; vascular system	Giving and receiving love; compassion and friendliness	Cold and distant.	Selfish and/or suffocating love.	Green

#	Center	Location	Gland	Influences	Functions	Underactive	Overactive	Associated Color
3	"Navel"[15]	Epigastrium	Pancreas	Stomach; Liver; Gall bladder; Nervous system; Vision	Action, initiative, self-esteem	Timid; passive and indecisive.	Aggressive, domineering.	Yellow
2	Sacral	Hypogastrium	Gonads	Reproductive System	Sensuality; feelings in general	Guilt, stiff or unemotional, not open to other people.	Over-emotional, high libido, too attached to people.	Orange
1	Base or root	Apex of sacrum or within the perineum	Adrenals	Kidneys; Spine; Motor function; Will	Self-identity	Fear, nervousness	Materialistic, greedy; obsessed with being secure and resistant to change.	Red
	Spleen[16]	Left hypochondriac region	Spleen	Blood; Lymph; Skin				
	Alta Major	Base of occiput	Carotid plexus	Blood pressure; Spine				

---

[15] Not only is it *not* located at the navel, it is also often called the "solar plexus" chakra, an erroneous and misleading name.

[16] The sacral chakra is sometimes called the "spleen" chakra, which is cause for some confusion.

Fourth-tier Assessment

The medical history and the sequence of events in the patent's life are gathered and put into a sequence from current to most distant. "What happened before that?" may have to be asked many times in order to get a complete picture.

Organic and environmental factors are noted as well.

Summary

Adding to the knowledge regarding the patient gathered at the outset, one adds the correlates of subtle or energetic disorders, the timeline of trauma or difficulties, and the somatic problems encountered along the way. Here is an example:

The patient is a 19 year-old woman who consults you for anxiety and sleeplessness. She began having problems sleeping several months ago.

First-tier assessment: Slightly disheveled, normal motor ability, soft speech, cooperative but conversationally inhibited; anxious mood, sad affect; alert and oriented. She does not initiate conversation and has poor eye contact.

Second-tier assessment: Asked about frustrated ambitions, she becomes more talkative. She is unable to fully indulge in her artwork, as her family insists that she get a more "practical" education. Instead of going to art school as she would like, she is in business school. She becomes more conversant when she talks about her oil painting.

When asked about other conflicts, she says that people don't hear her. She finds it not only difficult to speak her mind with her family but also with friends. When she does voice her opinions, she says, they are dismissed.

Third-tier assessment: Consulting the foregoing chart, you see that the likely impacted energy center is the throat chakra, from multiple perspectives: hypothyroidism, limitations of self-expression, shyness, and a history of respiratory problems as well.

Fourth-tier assessment: You make note that she has a history of recurring bronchitis and has recently recovered from another episode. Medications include Rx Ativan for the anxiety and insomnia (which has not completely relieved either), and Rx Synthroid for hypothyroidism.

An initial treatment plan would include a botanical medicine to lessen the anxiety and insomnia. She would be given some dietary guidelines based on these two problems, and possibly some supplemental nutrition if there are clear indications of deficiency. She is advised to receive some counseling apart from somatic therapies, because it is important that she have an open ear for her problems. Some somatic therapy sessions are scheduled, in which she receives acupuncture according to the assessment, combined with light therapy. The associated color with the throat chakra is blue, so it is possible that blue light (applied particularly to the throat area) could be an initial part of the treatment plan, changing the color as more information is available (probably orange, as we will see in Chapter Five). As she improves, or as her symptoms change—as is often the case—the more specific individualizing symptoms will point to a specific homeopathic medicine. This can provide yet another force for establishing balance on a deep level. It may be given in addition to a homeopathic *nosode*, or highly diluted disease agent[17].

Although this is skipping ahead somewhat, the chapter on assessment must logically include some idea of what will be done with the information once it is gathered. Mental health professionals reading this will find a decidedly different approach than the usual one of matching up the patient with the pathology.

---

[17] In this case, testing would reveal if the body's memory of an organism such as *pneumococcus* or *coxsackie* (from previous bouts of bronchitis) were having an ongoing influence on her health. Neutralizing them might be essential for fully clearing the emotional problems.

The PENE Model

It should be obvious by now that multiple mechanisms are being considered in this method of assessing psychoemotional problems. We are looking at what might be called a psycho-emotional-neuro-endocrine model, which we will dub the PENE model. Without taking into account the subtle perineural system that influences the nervous system, which in turn influences hormonal balance, much is lost and the best one can hope for is to slowly resolve problems with years of psychotherapy, or damp down symptoms with drugs.

From this point, we may now explore the treatment modalities in detail, and their effects on both gross and subtle physiology.

4

NUTRITION: DIET and ORTHOMOLECULAR THERAPY

We will first give an overview of the use of general dietary guidelines and nutritional supplements in the treatment and prevention of psychoemotional problems. Following this, we include more specific protocols for several common disorders.

Dietary Therapy

That diet has an impact on mental and emotional states is incontestable. While much of the knowledge concerning this has been with us for decades, this was practically a non-issue in conventional mental health circles, where such suggestions as diet playing a role in psychoemotional disorders were associated with radical "fringe" practitioners and "quack" therapies.

The dismissive attitude began to unravel over time with more scientific scrutiny being applied to the role nutrition plays in health and disease. Today, even the most conservative health care practitioner would not dare to completely dismiss diet as both a causative factor and as a therapeutic tool (at least in some conditions).

What we know from the research of recent years has not brought to light many errors concerning the nutritional teachings in the field of natural medicine from fifty years ago. Terminology may change in some cases—what we once called "Vitamin F" is now known as "essential fatty acids", for example—but while we know more about *why* things work as they do, the recommendations of those seemingly "unorthodox" experts of years past remain valid.

At different points in this book we discuss the role of neurotransmitters in the brain and their relevance to mental and emotional states. These brain chemicals are stimulated by, and often contained in, foods. Notable examples of these foods touch on another subject considered in Chapter Eleven: "glandulars" (in traditional parlance). They are organ tissues consumed as food—or as supplements—for their ability to supply the same organ or tissue with needed elements for healing.

It is no surprise that naturopathic doctors years ago were recommending wheat germ, oatmeal, granola, cottage cheese, peanuts and nuts (especially cashews), turkey, and fish for psychoemotional disorders. These foods are rich in tryptophan, an essential amino acid that is a precursor to serotonin in the brain.

We now also know that beans, fish, nuts, and red meat are rich in glutamic acid, from which our bodies can create glutamine. It is one of the amino acids formed within the human body, and which improves mental functioning, modulates blood sugar levels, and maintains muscle mass.

What we are saying is that before the commonly accepted model of brain neurotransmitter dysfunction being the "cause" of psychiatric disorders, foods were being prescribed in natural medicine that now have explained mechanisms of action in light of modern science. And while observations of drug actions on neurotransmitters are prized today, few studies of *food* effects are given notice[18].

---

[18] An exception: Salomon, R., Mazure, C., Delgado, P., Mendia, P., Charmey, D.; Serotonin Function in Aggression: The Effect of Acute Plasma Tryptophan Depletion in Aggressive Patients, *Biological Psychiatry* 35 (8): 570-572

For most of the more common neuropsychiatric problems we see in daily practice, there are some broad dietary guidelines that apply:

- **Sugar consumption must be reduced.** Sucrose is a trigger for many of these conditions[19], and the ubiquitous presence of high-fructose corn syrup (HFCS) in processed foods today is contributing to mental as well as physical illness. Manufacturers even add this sweetener to foods that are not confectionary items, like saltine crackers. Craving for sweets is typical in both pre-depressive and depressive states. The average patient will have reduced appetite and/or irregular eating times, leading to vitamin deficiencies and hypoglycemic episodes, which in turn cause more sweet cravings. Eating frequent small meals is a good policy, in order to avoid blood sugar drops.
- **A nutrient-rich diet must be followed**, in order to offset the deficiencies caused by meals consisting of processed, overcooked, or simply nutritionally empty foods. In particular, mineral reserves are typically depleted by the time someone displays symptoms. Minerals must be consumed and absorbed in order to replenish the essential alkaline buffers that have been lost. In particular, magnesium levels are critical for emotional balance and deficiencies of this mineral are linked to anxiety and depression.

---

[19] Westover, A.N., Marangell, L.B., A Cross-National Relationship Between Sugar Consumpotion and Depression? *Depression and Anxiety* 2002; 16:118-120

- **Water intake must be optimal.** By this we mean a steady supply of good quality water, at regular intervals throughout the day, to prevent dehydration. This condition is often present in patients seen for common mental and emotional disorders[20]. It is especially important to re-hydrate patients who are on medications that deplete the body of fluid (and essential minerals). Filtered water is preferred to tap water; distilled water is not recommended, as it will remove even more minerals.
- **Caffeine consumption should be modified.** It is typical in emotional trauma, cognitive problems, depression, and even in anxiety to increase coffee consumption. Medicating with caffeinated beverages in order to function is a stopgap measure that becomes routine. It leads to dehydration and has been linked to an exacerbation of stress reactions and anxiety, insomnia, and depression. Although it has mood-elevating effects that relieve depression, higher amounts of coffee consumption will have the opposite effect and increase depression, likely through a depletion of Vitamin B6. The problem is that the amount needed to do this is different for each person. A depressed person can easily cross the line from short-term help to constant worsening of his condition.
- **Food intolerances should be avoided.** This is not to say that allergy testing should be done to find which foods the person may be allergic to. Food intolerances are not always an immunoglobulin E (IgE) reaction, nor an immunoglobulin G (IgG) reaction, two common immune system responses. Sometimes perfectly healthy foods affect a person in a negative way that subtly undermines that person's total health[21]. *There need only be one food*

---

[20] Standing Committee on the Scientific Evaluation of Dietary Reference Intakes, "Dietary Reference Intakes: Water, Potassium, Sodium, Chloride, and Sulfate". *Institute of Medicine, Panel on Dietary Reference Intakes for Electrolytes and Water*, Feb. 11, 2004

[21] King, D., Mandell, M.; "A Double Blind Study of Allergic Cerebro-viscero-somatic Malfunctions Evoked by provocative Sublingual Challenges with Allergen Extracts", *Proc. 12th Advanced Seminarin Clinical Ecology*, Key Biscayne, FL, 1978.

trigger, eaten on a regular basis, to keep a person in a pattern of dysfunction. There are various ways of testing reactions to foods, from kinesiological tests to electrodermal screening to live blood cell analysis.

Orthomolecular Therapy

The term *orthomolecular* has been applied now for decades to the practice of using nutritional supplements to prevent or reverse illness. While it became associated with those using large doses of vitamins for a drug-like effect, we use the term for its intrinsic meaning. As the word implies, this type of therapy corrects nutritional imbalances on a molecular level.

The choice of *what* to eat provides much in the way of raw materials—fats, proteins, carbohydrates, fiber, etc. Orthomolecular efforts, by our usage, concentrate on those micronutrients that may or may not be missing in the raw materials—vitamins and minerals in particular.

Nutritional deficiencies, often at a subtle level, can cause mental as well as physical illness. One does not have to have a full-blown case of beri-beri in order to have a deficiency disease. Trace elements necessary for normal organic functioning—yes, even brain functions—can be appallingly low, and yet the person can appear to be, if not vibrantly healthy, at least disease-free.

Even if a person's psychoemotional problems were not caused by nutritional deficiencies, it is still likely that they can be *helped* by nutritional intervention. When a healthy diet is being eaten, which should under normal circumstances provide everything for normal functioning, a pattern of dysfunction can still persist, until acted upon by the concentrated nutritional uptake provided by a regimen of supplements. And all too often today, because of modern agricultural practices, the foods chosen for their healthy nature do not contain all the nutrients they once did. Many people who are carefully choosing their foodstuffs are not getting what they believe they are getting.

Likewise, the nutritional supplements recommended by many authorities do not deliver as well. Orthodox physicians, as well as modern naturopathic doctors, make supplement recommendations for the treatment of emotional problems using what seems to be solid science. For example, there is now widespread agreement on both sides of the medical fence as to the benefits of omega-3 fish oils.

Mental and emotional conditions are responsive, the literature says, to supplemental magnesium, Vitamin B complex, and Vitamins D, A, C, and E. Yet, the products usually given to remedy these conditions contain synthetic vitamin analogues, isolates, and inorganic minerals with poor absorption. This can be a problem. While these materials are officially listed as nutrients in the United States Pharmacopoeia (USP), they are not necessarily the best choice for treatment.

Magnesium has shown to have a demonstrable impact on depression, even severe depression[22], but magnesium salts, not food-grade magnesium, have been used, and in amounts that constitute more of a drug action than a nutritive one.

The B vitamins have long been held to be critical for emotional health, and their deficiency has been shown to initiate anxiety, depression, and even epilepsy. Their importance in psychiatric treatment has been the subject of scrutiny in studies in Germany[23], where such topics are given a bit more serious consideration. Such studies always point out, for example, the specific actions of Vitamin B6 on neurotransmitters, or Vitamin B1 in nerve function. But when such supplementation is recommended, the more astute authorities advise adding a B-complex vitamin as well, due to the well-known effect of B vitamins taken in isolation causing deficiencies of *other* B

---

[22] Eby, G.A., Eby, K.L.; Rapid Recovery from Major Depression Using Magnesium Treatment, *Med. Hypothesis*, 2006;67 (2):515-527

[23] Hermann, W., Lorenzl, S., Obeid, R., Review of the Role of Hyperhomocysteinemia and B-vitamin Deficiency in Neurological and Psychiatric Disorders—Current Evidence and Preliminary Recommendations, *Fortschr Neurol Psychiatr.* 2007; 75(9): 515-527

vitamins. This begs the question, "Why not take them all to-gether in the same source, in the first place?" There is the feeling, in many circles, that nutrients are not effective unless they are treated like drugs. Or perhaps the researchers know their papers would not be published otherwise.

Vitamin D has received much attention in the last ten years for its health effects, particularly mental health. Using the serotonin model, this vitamin with dramatic benefits for "brain chemistry" has been lauded in prestigious medical journals[24]. Its successful application in depression, and particularly Seasonal Affective Disorder, has been studied and confirmed[25]. As has been dis-cussed in Chapter Five, the effects of light are not purely subtle and mysterious, or in the case of criticisms of chromotherapy, metaphysical in nature. The chemical responses within the body as Vitamin D is activated by sunlight are well understood. But how much more benefit if the supplementary Vitamin D was the easily assimilable food-derived D complex? To date, the promi-nent studies focus on the cholecalciferol form of Vitamin D3. This isolate, given in a dose over ten times the usual dose, caused the subjects to improve on all depression scale measure-ments. There was improved negative mood and enhanced already-positive mood, and to a degree greater than sunlight produced in the seasonally depressed subjects[26]. Well, why not? Even a nutrient, given in such a large dose, is going to have a drug effect. But is that a viable option for health maintenance? It is unfortunate that no study followed this one to find out the effects of Vitamins D1, D2, and D4 being displaced. One al-most wonders if researchers feel that nature made a mistake putting those factors in there in the first place.

---

[24] Wilkins, et al.; Vitamin D Deficiency is Associated With Low Mood and Worse Cognitive performance in Older Adults, *Am J Geriatr Psychiatry,* 2006: 12(12): 1032-40
Partonen, T., Vitamin D and Serotonin in Winter, *Med Hypotheses.* 1998; 51(3): 267-268
[25] Gloth, F.M., Alam, W., Hollis, B., Vitamin D vs. Broad Specrtum Pho-totherapy in the Treatment of Season Affective Disorder, *J Nutr Health Aging.* 1993; 3(1): 5-7
[26] Landsdowne, A.T., Provost, S.C., Vitamin D3 Enhances Mood in Healthy Subjects in Winter. *Psychopharmacology* (Berl). 1998; 135(4): 319-323

Plant-based Vitamin A consists of beta-carotene, retinyl esters, and mixed carotenoids. The common chemical analogues are Vitamin A acetate and Vitamin A palmitate. Supplements often contain just the isolated beta-carotene. Animal source Vitamin A is in the form of retinol. Retinol is converted to retinyl palmitate and retinaldehyde and retinoic acid, which are used often for cosmetic products (and the acne drug Accutane, which came under fire for causing birth defects).

Real Vitamin C is found in many fruits and vegetables, but it is actually nine different naturally occurring chemicals ("vitamers"), accompanied by many flavonoids (originally thought to be a single substance and called "Vitamin P"), each with their own actions. There are nine vitamers comprising natural Vitamin C and two dozen substances in the whole Vitamin C complex. One vitamer is ascorbic acid, which serves to protect the nutrients in the plant from oxidizing and losing their potency. It is a preservative and antioxidant for the plant. Most authorities talk as though ascorbic acid and Vitamin C were synonymous.

After consuming the food containing the Vitamin C complex, ascorbic acid is excreted in the urine. If ascorbic acid is "equivalent" to Vitamin C, why does it not stay in the body? Yet, commercially produced Vitamin C supplements are typically some form of isolated ascorbic acid.

And Vitamin E? The usual form recommended was found to increase risk of stroke by 113%[27].

Thankfully, omega-3 fatty acids found in supplements remain natural. Seeds, nuts, fish oil, other oils such as canola oil and olive oil are rich sources. Even some vegetables as broccoli, cauliflower, and bok choy contain them.

With the arrival of scientific acceptance of naturopathic concepts of nutrition came the insistence that "natural vitamins are no bet-

---

[27] *Stroke,* 2004; 35: 1908-1913

ter than synthetic or inorganic vitamins." The scientific literature is full of statements about their identical molecular structure, and the inability of the body to distinguish a synthetic from a natural vitamin. The new generation of practitioners, with science endorsing at least some of their concepts, has gleefully accepted this dictum. This is also easy because the manufacturers of such supplements endow their schools and support the profession. The small number of companies that manufacture natural vitamins and minerals have no such presence.

It is put to the reader that natural supplements made from whole food concentrates are simply better. The nutrients are more bioavailable, and the full spectrum of known and unknown elements are there when the nutrient is presented in its naturally occurring complex, rather than when isolated. While many on a synthetic vitamin regimen have enjoyed improved health (or at least relief of their symptoms), the difference with whole food supplements is quite obvious.

USP nutrients are useful and certainly have their positive effects. They are valuable for the short-term treatment of acute illnesses and in severe deficiency conditions, especially when full-spectrum nutrients are not available (medical treatment in third-world countries, etc.) But isolated nutrients are essentially unnatural because deficiencies are never singular. A true nutritional deficiency means an entire complex is missing, which can never really be filled with a synthetic isolate.

How, then, are they useful? Applying USP vitamins and minerals according to the symptoms or conditions at hand is using a supplement like a drug. Synthetic isolates, given in therapeutic doses, create a drug effect. Symptoms are modified fairly quickly. Like drugs, however, there is a flip side: An underlying process may be unchanged, and there may be side effects. This is particularly true of the "megadose" form of therapy, which was the methodology that gave rise to the term "ortho-molecular" in the first place.

Synthetic vitamins are *analogues* to the natural vitamins; they are chemically similar but not the same. Their effects differ because of this difference, but also because all the other co-factors are absent in an isolate. Therefore, while they may be useful for stabilizing the patient and reducing symptoms as a short-term measure, continued use will not elevate the person to a higher state of health. While natural supplements can also be used in acute stages, their long-term use is responsible for actual resolution of the problems.

Bibliography

- Barborka, C., *Treatment By Diet,* 2nd Edition, Lippincott, 1935
- Bicknell, F., Prescott, F., *The Vitamins In Medicine,* 3rd Edition, Lee Foundation for Nutritional Research, Milwaukee, 1953
- Bieler, H., *Food Is Your Best Medicine,* Random House, 1965
- Jensen, B., *Foods That Heal,* Avery Pub. 1988
- Lee, R., Stoltzoff, J., *The Special Nutritional Qualities of Natural Foods,* Lee Foundation for Nutritional Research, Milwaukee, 1942
- Thiel, R., Fowkes, S., *Can Cognitive Deterioration Associated With Down Syndrome Be Reduced?* Medical Hypotheses, 2005; 64(3)
- Thiel, R. *Natural Vitamins May Be Superior To Synthetic Ones.* Medical Hypotheses, 2000 55(6)

5

PHOTOTHERAPIES

Phototherapy is the use of light, in both its invisible and visible forms. It has been a modality in natural medicine since the early days. Arnold Rikli (1823-1926) of Austria established this as a treatment in the days when Naturopathy was not yet an organized school. He established the first sanitarium using the "light and air cure", at Veldes-Krainola, in 1848. Later (and more famous) conventional physicians such as Finsen and Rollier used light therapy in the treatment of tuberculosis and other diseases at their sanitariums, and the practice spread worldwide. It is typically forgotten in the age of pharmaceutical-dominant medicine that orthodox medical facilities used many natural therapies quite effectively. In recent years, there has been a bit of a resurgence of scientific validation of phototherapy.

By now, most people have heard of the use of high quality light in the treatment of the depressive complaint known as Seasonal Affective Disorder (SAD). Once something becomes commonplace, it is not questioned; but the history of using light to treat maladies has been marked by controversy and outright hostility.

Full-spectrum Lighting

Photobiology, the study of the effects of light on living things, would probably not have achieved the scientific footing it has if not for the work of John Ash Ott. In recent decades, it has become a matter of widespread knowledge in the scientific community that the quality of artificial light has a negative impact on the human organism.

John Ott was hired by Disney Studios to create the proper lighting conditions for time-lapse photography of plants. He studied the responses of flora and fauna to specific wavelengths of light. By filming through a microscope, he discovered that pigment granules in both plants and animal cells exhibited changes in behavior when exposed to different parts of the light spectrum. He also found that sickness rates were dramatically reduced in workplaces where the windows admitted ultraviolet light[28]. Ott realized that the deliberate screening of parts of the light spectrum with sunglasses, window glass, tinted windows, suntan lotions, etc., we were creating a new illness—"malillumination". It was the environmental equivalent of malnutrition due to lack of trace elements in the diet.

Ott published his first book, *Health and Light*, in 1973 and pioneered the production of artificial lighting that mimicked the qualities of sunlight. "Full-spectrum lighting" entered the marketplace and has been a controversial topic, despite research at Johns Hopkins University that showed it to be clearly of value in treating Seasonal Affective Disorder. Accredited investigators have mapped out how light of a certain quality activates responses from the pineal and pituitary glands in the brain, but the medical community has been negative about validating the concept of sunlight curing disorders. Frustrated, Ott established the Environmental Health and Light Research Institute in Florida, where the effects of light on a number of conditions have been studied. A correlation between lighting and hyperactivity in school children has been made, for example. Even light's effects on tumor growth have been observed.

Light has been shown to be an effective treatment for seasonal and non-seasonal depression as well as bipolar depression, pre- and postpartum depression, and insomnia and circadian sleep disorders. One primary mechanism appears to be the stimulation of the pineal gland by way of the optic nerve[29].

---

[28] Ordinary window glass filters out UV light, which is why you cannot get a suntan indoors.

[29] Altschule, M.; *Frontiers of Pineal Physiology*, 1975 M.I.T. Press

Recent research on Langerhans cells found in human skin makes Ott speculate that the cells are actually biological solar cells, passing solar energy into the body.

The Mysterious Effects of Color

A different aspect of light will be discussed next. Three examples will introduce the subject:

An FDA-approved device for treating acne is currently on the market and is nothing more than a lamp with red and blue lights, with all its action in the visible spectrum.

The "bili light" used in hospitals for treating jaundiced infants is quite old and still in use. It uses filaments producing blue light in a specific range (425-475 nm) that causes excretion of bile salts in the body.

 In 2009, a curious discovery appeared in a peer-reviewed medical journal: Blue light was able to kill two strains of methicillin-resistant staphylococcus aureus (MRSA), the most ubiquitous and deadly antibiotic–resistant bacterial infection[30]. The "photoirradiation" used in the study is significant for our discussion here, because it used light in the 470-nm range, which emits no ultraviolet radiation. The effects, therefore, were not the usual chemical effects from actinic rays, associated with UV light. Actinic rays have known antimicrobial effects, but this was a very different matter.

Why would these examples be noteworthy? A type of phototherapy used in natural medicine for decades (and highly contested by the medical establishment) is *chromotherapy*.

---

[30] Enwemeka, C., Williams, D., Enwemeka, S., Hollosi, S., Yens, D.; Blue 470-nm Light Kills Methicillin-Resistant Staphylococcus Aureus (MRSA) in Vitro, *Photomed Laser Surg* 2009 Apr; 27(2): 221-6

Chromotherapy

Chromotherapy, treatment with colored light, has been a political "hot potato" in medicine since the 1920s. Each color was found to have predictable effects on certain tissues in the body, and the ability to resolve certain disease conditions. The discovery that frequencies in the visible light spectrum have therapeutic actions, just as those in the invisible parts of the spectrum (infrared, ultraviolet, etc.) was extremely unwelcome as the large drug houses were establishing their dominance over medicine.

Because the ability to produce colored light can never be controlled or outlawed, and because it was being used by an increasing number of trained doctors, the method was strongly disparaged. Equipment for treating was confiscated from physicians' offices and from private homes, and for decades after, any thinking person would be disinclined to argue if told that shining colored light on the body to cure disease was quackery of the first rank.

Unfortunately, this had a chilling effect on naturopathic doctors using chromotherapy, who were increasingly inclined to use only those methods deemed "defensible". Not wanting to appear unscientific, the use of colored light declined over the years, and the experts in the field died off.

Two organized systems of chromotherapy were prominent: *Spectro-chrome*[31], originated by Dinshah Ghadiali, and *Harmono-chrome*, devised by Dr. Carl Loeb. The two systems had different types of equipment for generating the colored light.

Dr. Harry Riley Spitler was a naturopathic doctor and optometrist who took Loeb's course and worked for him for a time. Eventually, after editing a journal devoted to light therapy, Spitler founded the Central States College of Physiatrics, a naturopathic school in Eaton, Ohio (later in Columbus). Chro-

---

[31] Still the most widely used system of color therapy in the world.

motherapy was taught there. Spitler became famous for writing *Basic Naturopathy*, a major textbook in the 1940s and 1950s.

Dr. Spitler called his approach of using light to balance the human system "Syntonics". Because he was also an optometrist, Spitler did much research into the effects of colored light on the eye and on vision. Developing devices for the introduction of color to affect the optic nerve, he founded a postgraduate training curriculum based on it, the College of Syntonic Optometry, which is still in existence.

He was not alone in concentrating the effects of colored light on the optic nerve. Paul Wendel, N.D., in his textbook on bloodless surgery[32], advised the use of blue-tinted goggles in preparing patients for treatment. By shining light through the blue lenses for ten minutes, the patient becomes relaxed, and the lenses are changed to a color combination and further exposure to light is made. When the patient begins to see a yellow or gold shade, it is evidence that the pain centers in the brain have been sedated. The bloodless surgery procedure is then performed.

Why does chromotherapy work?

Every element, when placed in a spectroscope, gives off a distinctive identifying set of emission lines, color bands (called Fraunhofer lines) that are specific to that substance. When exposed to light, the element will absorb from the light those frequencies (colors) that it characteristically gives off. Therefore, full-spectrum white light (which contains all colors) acts as a "nutrient" for the element. Limiting the light to only the characteristic color intensifies the absorption of that color. Dinshah and Loeb both matched the Fraunhofer lines associated with a particular element to find the color related to its function. Then sick people's conditions that were associated with a dysfunction of that element in the body were treated with that color, and observed for changes. This has led to a large repository of

---

[32] *Naturopathic Bloodless Surgery with Technique and Treatments,* by Paul Wendel, N.D., self-published, 1945

knowledge regarding which colors have been found to act positively on which illnesses. Spitler did the same but concentrated on application through the eyes to create neurological changes via the brain. Spitler theorized that the stimulation of either the sympathetic or parasympathetic nerves by the colored light was the mechanism by which his chromotherapy worked, since illness was seen to be a disequilibrium of the autonomic nervous system.

Today, it is well established that visual stimulation creates the "entrainment" response in the brain, and habitual patterns can be offset by application of bursts of color along the optic nerve, thus interrupting those patterns and helping to establish new ones. For this reason, chromotherapy has been applied by Syntonic optometrists with great success in emotional problems and information processing disorders.

As a general theory it is accepted that in disease, the structural patterns of the neural pathways are abnormally aligned. This can be no truer than in the case of mental and emotional disorders. Chromotherapy appears to release chemicals that accomplish a re-alignment of those pathways with enhanced nerve cell growth. On a more broad level, mal-illumination (a condition resulting from missing frequencies of natural light, due to artificial lighting or insufficient exposure to sunlight) can be a causative factor in disease and can be corrected by supplying exposure to the colors the person is lacking.

Dinshah's theory was that, since spectroscopy proves that each element emanates its own specific frequency of light (a sort of "vibrational fingerprint"), illness produces changes in those frequencies in different tissues. Disharmony between the various systems in the body with their composite frequencies would be a disease state, which could then be manipulated with the needed frequencies of light (colors), which would return resonance to the abnormal, out-of-phase state at the molecular level. Matter being simply condensed light, as it is now referred to in modern physics, it follows that action on matter by use of colored light is not such a metaphysical proposition as formerly thought. At any rate, many decades of successful use with thousands of case histories on record confirm chromotherapy's place in the natural armamentarium.

Chromotherapy in Emotional Disorders

The recommended procedure is to apply the appropriate colored light to the person's face and head while lying quietly in a dark room, ideally for one hour's duration. Treatment is most effective if done daily until better; however, twice to three times weekly in a clinic setting may suffice. The great beauty of chromotherapy is that the patient may be able to reproduce the treatment at home using very simple materials.

The most often-used protocols are given in this following brief list. While certain colors are frequently employed for specific conditions, the individual may most need a color that is not ordinarily associated with that problem. By consulting the next table, the individualizing actions of the colors can be matched up with the patient's medical history. The one that is the most applicable to the physical complaints has a chance of being most effective for the mental sphere also, despite that color's not being recommended for that particular emotion, etc.

Finally, included at the end of this chapter is a questionnaire that can be used to better determine the color that the patient is requiring most in his or her "color starvation".

Quick-reference Protocols

Adrenal Dysfunction contributing to symptoms:
- **Lemon** anterior thorax for systemic treatment
- **Scarlet** over kidneys
- **Green** on face to affect pituitary

Alcoholism:
Acute:
- **Blue** anterior thorax if relaxation needed
- **Magenta** anterior thorax
- **Scarlet** anterior thorax if collapse is possible (will raise blood pressure)
- **Orange** on throat and chest if respiration rate is low

Chronic:
- **Magenta** anterior thorax
- **Red** on liver and stomach regions

Alzheimer's Disease:
- **Lemon** on face and anterior thorax
- **Magenta** on face, throat, chest and kidneys

Anxiety:
- **Violet** on face
- **Purple** on face, throat and chest if pulse fast or throbbing

Appetite, Decreased:
- **Green** or **magenta** on anterior thorax and face
- **Yellow** to abdomen may help

Appetite, Increased:
- **Lemon** on face and anterior thorax for systemic treatment
- **Indigo** on face and anterior thorax if bulimia/purging

Attention Deficit Hyperactivity Disorder:
- **Magenta** anterior thorax
- **Green** on face and back of head

Bipolar Affective Disorder:
- **Purple** or **indigo** on head and face during manic phase
- **Yellow** on head and face during depressive phase

Cognitive Difficulties, Mental Fatigue:
- **Green** on face and anterior thorax to balance pituitary

Dementia, Non-Alzheimer's:
- **Green** on face and anterior thorax
- **Magenta** on face, throat, chest and kidneys

Depression:
- **Violet** on face/head for acute
- **Purple** on face/head if agitated
- **Magenta** on face/head balances emotions

Fatigue:
- **Lemon** on anterior of body
- **Indigo** on liver as a secondary treatment

Hyperthyroidism:
- **Indigo** on anterior head and throat
- **Green** acts on pituitary and may be useful
- **Purple** on anterior head and throat if high blood pressure or exophthalmic

Hypothyroidism:
- **Green** on anterior head can balance pituitary
- **Orange** on anterior throat
- **Indigo** on throat to reduce goiter
- **Scarlet** on entire body if adrenals are low

Insomnia:
- **Violet** on face
- **Purple** on face, if tachycardic or agitated

Premenstrual Syndrome:
- **Green** on head, anterior and posterior
- **Magenta** on head if emotions are erratic

Tourette's Syndrome:
- **Violet** and **purple** on face/head

Actions of the Colors

RED Stimulant for the nervous system and liver; tonic; builds blood (platelets, hemoglobin, etc.); stimulates skin to expel toxins; counter-irritant for burns.	**BLUE** Mild sedative; anodyne; antipruritic; antipyretic; diaphoretic; stimulates pineal gland.
ORANGE Stimulates respiratory system, stomach, bone growth, thyroid gland; depresses parathyroid; antispasmodic; galactagogue.	**INDIGO** Stimulates parathyroid; Depresses thyroid, respiratory system, mammary glands; astringent; hemostatic; increases phagocytosis.
YELLOW Stimulates nerve tissue, lymphatic system, pancreas, intestines; increases production of digestive fluids; depresses the spleen; anthelmintic.	**VIOLET** Tranquilizer; stimulates spleen; increases leukocytosis; depresses muscles, lymphatic system, pancreas, nervous system.
LEMON Brain stimulant; stimulates bone growth; promotes tissue repair and nutritional uptake; thrombolytic; mucolytic; used in chronic disorders.	**PURPLE** Emotional sedative/relaxant, soporific; depresses kidneys and adrenals; decreases sensitivity to pain; lowers heart rate and blood pressure; lowers body temperature; increases tone of veins; vasodilator.
GREEN Equilibrator of brain; stimulates pituitary; stimulates muscle and connective tissue repair; antiseptic/disinfectant/antimicrobial.	**MAGENTA** Balances: emotions, heart, blood vessels, kidneys, adrenals, reproductive organs.
TURQUOISE Brain depressant; skin tonic; antipyretic; used in acute illnesses as a rapid alterative.	**SCARLET** General stimulant; stimulates kidneys and adrenals; vasoconstrictor; increases heart rate; stimulates reproductive organs

Questionnaire for Determining Color Choice

The scores should reflect somatic characteristics as well, even if the goal is to balance the patient mentally/emotionally. The medical history and physical symptoms are essential, for they are concurrent, if not contributing, factors in the psychoemotional illness of the person.

RED

0 = Never   1= Not often   2= Sometimes   3=Often
4=Very often   5=Constant

#	Symptom
	Amenorrhea (absent or late menses)
	Anemia
	Appetite low
	Arthritis
	Cold extremities
	Colds and flu often
	Color preference is blue
	Fatigues easily
	Food: craves salty foods, prefers warm food to cool
	Digestion poor
	Moody disposition
	Overweight
	Pale skin
	Quiet personality
	Sex drive or performance low
	Stools often loose
	Total

SCARLET

0 = Never   1= Not often   2= Sometimes   3=Often
4=Very often   5=Constant

#	Symptom
	Amenorrhea (absent or late menses)
	Apathy
	Activity aggravates symptoms
	Blood pressure tends to be low
	Cold extremities
	Color preference is purple or lavender
	Depressed
	Eats often
	Enjoys using mind/studious
	Eyes puffy
	Fatigue in afternoon
	Feet or ankles swell
	Food: craves sour flavors
	Irritable when hungry
	Intuitive
	Shaky if hungry
	Skin dry
	Skin eruptions
	Sleepy after meals
	Stiff joints on arising
	Urinates frequently
	Total

ORANGE

0 = Never   1= Not often   2= Sometimes   3=Often
4=Very often   5=Constant

#	Symptom
	Asthma
	Belching
	Blackheads
	Bloated abdomen
	Color preference is indigo or dark blues
	Constipation
	Craves tobacco or enjoys smelling it
	Depressed
	Flatulence
	Food: craves pungent foods; garlic, onions, spicy, etc.
	Hair and scalp dry
	Hiccups
	Indigestion
	Menstrual cramps
	Muscle cramps, esp. at night
	Neck stiff
	Osteoarthritis
	Osteoporosis
	Short of breath on exertion
	Skin dry
	Stools dry
	Thyroid: low function
	Water: dislikes drinking
	Total

YELLOW
0 = Never   1= Not often   2= Sometimes   3=Often
4=Very often   5=Constant

#	Symptom
	Appetite poor
	Arthritis
	Belching, sour
	Burning in esophagus
	Color preference is gray or violet
	Constipation
	Diarrhea
	Fatigue from nervous strain
	Fear of being wrong
	Food: picky eater
	Heartburn
	Heel spurs
	Personality is stubborn or strict, dogmatic
	Stools have mucus or fat with them
	Total

LEMON

0 = Never   1= Not often   2= Sometimes   3=Often
4=Very often   5=Constant

#	Symptom
	Bloated abdomen
	Breath smells foul
	Bruises easily
	Colds and flu often
	Color preference is gray or black
	Constipation
	Diabetes
	Drowsy often
	Fatigue/exhaustion from nervous strain
	Food: craves cheese
	Food sensitivities/allergies
	Eats often
	Memory poor
	Mucus overproduction
	Obesity
	Prostate enlarged
	Skin dry, lacks luster, slow to heal
	Skin eruptions
	Swelling of fingers or toes
	Total

GREEN

0 = Never   1= Not often   2= Sometimes   3=Often
4=Very often   5=Constant

#	Symptom
	Acne
	Abdomen protrudes; "pot belly"
	Body odor
	Colds and flu often
	Color preference is brown or tan
	Drug or substance abuse
	Food: fatty or oily foods are poorly tolerated
	Hair and scalp oily
	Muscles weak or flabby
	Nausea
	Personality: impulsive, dislikes being with people
	Sinus problems
	Skin eruptions: pustules
	Skin heals slowly or poorly
	Skin oily
	Underweight; weight gain is difficult
	Vaccination reactions
	Total

TURQUOISE

0 = Never   1= Not often   2= Sometimes   3=Often
4=Very often   5=Constant

#	Symptom
	Allergies
	Athlete's foot or other fungal infections
	Candida (yeast) infections
	Colds recur often
	Color preference is maroon
	Dental problems/gum disease
	Eczema, moist/weeping
	Facial skin unhealthy
	Foot odor
	Food: craves sweets
	Headaches
	Mouth ulcers
	Muscles weak or flabby
	Nails have ridges, are brittle, or weak
	Obesity
	Personality: skeptical, critical
	Skin flabby, prematurely wrinkled
	Skin itching
	Tremors
	Total

BLUE

0 = Never   1= Not often   2= Sometimes   3=Often
4=Very often   5=Constant

#	Symptom
	Angers easily
	Anxious
	Backaches
	Balding, premature
	Complexion red/ruddy
	Dandruff
	Ear infections
	Eyes red or burning
	Eyelids twitch
	Headaches, bursting or splitting type
	High blood pressure
	Hysterical sadness or anger; powerful emotions
	Migraines
	Nerve pain/neuralgia
	Nervous tics
	Pains are shooting, erratic
	Personality: easily upset
	Psoriasis
	Sciatica
	Sore throats
	Total

INDIGO

0 = Never   1= Not often   2= Sometimes   3=Often
4=Very often   5=Constant

#	Symptom
	Color preference is orange or yellow
	Eyes: cataracts
	Falling hair
	Food: prefers hot, spicy foods
	Food: prefers cold drinks
	Hot flashes
	Hyperactive
	Nosebleeds
	Personality: impulsive, aggressive
	Perspires easily
	Sense of smell poor
	Sex drive strong
	Sleeps poorly
	Throat scratchy
	Total

VIOLET

0 = Never   1= Not often   2= Sometimes   3=Often
4=Very often   5=Constant

#	Symptom
	Appetite strong; overeats
	Allergies (inhalant)
	Awkwardness, clumsiness
	Falling hair
	Color preference is white or green
	Dandruff
	Eats when nervous
	Eczema, dry
	Food: craves salty foods
	Metabolism slow; gains weight easily
	Numbness
	Obesity
	Personality is withdrawn, introverted, avoids responsibilities
	Psoriasis
	Skin tone poor; flaky skin diseases like seborrhea
	Stiffness
	Talks with hands, gesticulates
	Yawns frequently
	Total

PURPLE

0 = Never 1= Not often 2= Sometimes 3=Often
4=Very often 5=Constant

#	Symptom
	Argumentative
	Awakens easily; poor sleeper
	Dandruff
	Color preference is black
	Epilepsy
	Food: picky eater
	Headaches, pounding or pulsating type
	High blood pressure
	Menstrual flow excessive or painful
	Migraines
	Nervous tension headaches
	Night sweats
	Nightmares
	Personality is dramatic, confrontational, insecure
	Sex drive increased
	Sinus problems
	Skin dry, flaky
	Worries easily
	Total

MAGENTA

0 = Never   1= Not often   2= Sometimes   3=Often
4=Very often   5=Constant

#	Symptom
	Arthritis
	Chest feels tight on exertion
	Chest pains, dull
	Craves alcohol
	Dizziness
	Extremities swollen, puffy
	Eye: cataracts or glaucoma
	Facial swellings: around eyes, jaws
	Food: prefers fatty or rich foods, meat
	Heart: irregular beats
	Heart: history of heart attack
	Hemorrhoids
	Herpes eruptions
	High cholesterol or triglycerides
	Leucorrhoea ("the whites") vaginal discharge
	Personality: over-emotional, exaggerated, opinionated
	Skin dry
	Stroke, history of
	Varicose veins
	Total

#1_____   #2_____
#3_____

The top three totals indicate the primary, secondary, and tertiary colors for therapy. The highest score tends to indicate the constitutional color. The second highest score is a complementary color to use, and the third can supplement them both.

Bibliography:

- Breiling, B., et al., *Light Years Ahead*, Celestial Arts 1996
- Cleaves, M., *Light Energy*, Rebman Co., Pub., 1904
- Compilation of articles and authors, *Color Healing*, Health Research Pub. Co., 1999
- Cordingly, E.W., *Principles and Practice of Naturopathy: A Compendium of Natural Healing*, 1925 (reprinted by Health Research)
- Dinshah, D., *Let There Be Light*, Dinshah Health Society, 2nd Edition 1995
- Dinshah, D., *Spectro-Chrome Guide*, Dinshah Health Society 1997
- Douglas, W., *Into the Light*, Second Opinion Publishing, 1993
- Kovacs, R., *Electrotherapy and Light therapy*, Lea & Febiger, 5th Edition 1947
- Lieberman, J., *Light, Medicine of the Future*, Bear & Co., 1991
- Loeb, C., *A Course in Specific Light Therapy*, Self-published, 1939
- McWilliams, C.H., *Photobiotics*, 1997
- Ott, J., *Health and Light*, Devin-Adair Co., 1973
- Ott, J., *Light, Radiation, And You*, Devin-Adair Co., 1982
- Spitler, H.R., Class notes, 1931 by N.M. Abbe, D.M., D.C.
- Spitler, H.R., *The Syntonic Principle, Its Relation to Health, and Ocular Problems*, College of Syntonic Optometry, 1941
- Wesson, V., and Levitt, A., *Light Therapy for Seasonal Affective Disorder, "Seasonal Affective Disorder and Beyond"*, American Psychiatric Press, 1998:45-89
- White, G. Starr, *The Natural Way, or My Work*, 1924

6

ACUPUNCTURE

The basic premise of acupuncture is well known by now; we will provide a cursory explanation here. The theory is that a circulating bioelectrical energy (called *qi* in Chinese) in the body constitutes a controlling influence on organic processes. This energy circulates in a network of conduits (called *meridians*) that loosely follow the distribution of nerves and blood vessels. The meridians pass through various levels of tissue, becoming more superficial as they reach the ends of the extremities. Along the surface of the skin, there are points that act like valves for the transmission of energy along that meridian. By stimulating the points, one can affect the flow of bioelectrical energy and thus change functioning along that pathway and also in the internal organ associated with it. The type of stimulation and choice of the appropriate point at the appropriate time determine the effects. Needling or otherwise stimulating the point in one way increases the transmission of energy; a different stimulation inhibits it.

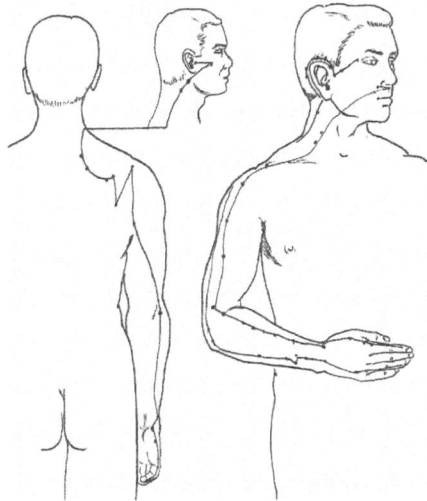

Part of the difficulty with the acceptance of acupuncture is this theory, which assumes anatomical connections that are not known to exist in modern physiological knowledge. However, the most recent research has shown that stimulating acupuncture points activates areas of the deep brain.

The usual method of inserting and then manipulating the needle creates a sensation known in Chinese as *deqi*. The characteristic sensation that is elicited when the acupuncture effect is beginning to take place is seen in classical acupuncture theory as a manifestation of the *qi* or bioelectrical energy flowing through the meridian. In traditional Chinese medicine, eliciting the *deqi* sensation is considered essential for efficacy.

Tests have been done to compare the traditional needling method with other stimulation of the skin (and the expectation that acupuncture was being performed). The *deqi* response was elicited in 71% of the true acupuncture procedures compared with 24% for tactile stimulation. So the actual manipulation of the acupuncture point appears to be causing the sensations, and not the simple belief that acupuncture is being done.

A research team of scientists from University College London, Southampton University, and the University of York, found that superficial or light needling (tonification technic, in acupuncture terminology) resulted in activation of the motor areas of the cortex. This is a normal reaction to pain. However, with deep needling (sedation technic), the limbic system, part of the pain matrix, is *deactivated*.

The researchers have mapped the effect of acupuncture on the brain. The study makes it easier to understand in scientific terms how acupuncture works. The findings, published in *Brain Research*, confirm that acupuncture has an effect on specific areas of the brain, and that the *deqi* sensation creates brain changes and is not an imaginary effect.

An earlier 2007 study exploring *deqi* and the effects of acupuncture reported, "The complex pattern of sensations in the *deqi* response suggests involvement of a wide spectrum of myelinated and unmyelinated nerve fibers, particularly the slower conducting fibers in the tendinomuscular layers."[33] The research authors added that the findings were "clinically relevant and consistent with modern concepts in neurophysiology."

With the new studies, scientists have taken things a step further by mapping the effect of acupuncture on specific neural areas in the brain.

Dr. Hugh MacPherson, of the Complementary Medicine Research Group in the Department of Health Sciences at the University of York, says: "These results provide objective scientific evidence that acupuncture has specific effects within the brain which hopefully will lead to a better understanding of how acupuncture works."

More recent studies at the University of New South Wales[34] have shown significant effects from acupuncture in the parts of the brain that regulate emotion. Scientists from the UNSW School of Psychiatry used functional magnetic resonance imaging (fMRI) to monitor changes in the prefrontal cortex and subcortical nuclei when acupuncture points were stimulated. Fiber optic lasers were used to stimulate the points, which made for an excellent blinding method, because the subjects could not feel if they were getting a real treatment or placebo. Study chief Prof. Perminder Sachdev said, "This is the first MRI study to find that laser stimulation of a suite of acupoints on the body in healthy individuals produces changes in brain regions that may be relevant to treating conditions such as depression."

Lead author Dr. Im Quah-Smith noted that "The most consistently reported finding in antidepressant treatments is that they lead to a normalization of activity in the prefrontal cortex, with additional changes in the limbic cortex and the frontal lobe." The ability of laser acupuncture to affect these structures shows

---

[33] Hui, K.S., et al.; Characterization of the "*Deqi*" Response in Acupuncture, *BMC Complementary and Alternative Medicine* 2007 (10) 7:33
[34] http://www.unsw.edu.au/news/pad/articles/2010/sep/Acupuncture

promise that future trials using actually depressed subjects will again affirm the utility of acupuncture.

Acupuncture in Psychoemotional Disorders

The subject of this book is the use of natural treatments for mental and emotional disorders, and acupuncture is a prominent tool in the armamentarium of those wishing to provide such treatment. It is therefore important to point out that incorporating acupuncture in a treatment plan is consistent with the findings of modern science. Acupuncture not only affects the brain, but in fact that very mechanism is (at least partly) responsible for creating the bodily effects observed for thousands of years. Acupuncture's age-old theories, allusions to dragons and wind invading the body, etc., no longer have to be seen as an embarrassment that questions the validity of the practice. Modern science has spoken: Acupuncture causes distinct changes in brain chemistry. As has been pointed out here, the conventional view of mental and emotional illness has been increasingly explained in terms of "changes in brain chemistry". Given this, one might think that there would be an increasing embracement of acupuncture as a treatment modality.

But the major funded studies usually focus on pain relief. This is partly understandable, since this benefit is what most people associate with acupuncture. But acupuncture treats far more than painful conditions. The medical and pharmaceutical industry is not anxious to portray an ancient treatment from a pre-scientific era as a viable treatment for organic disease. And with antidepressant drugs enjoying booming sales worldwide, its use in this realm is certainly unwanted.

One trial[35] compared acupuncture and expectation (placebo response) to antidepressant medication. Critics of acupuncture were quick to point out that there was no significant difference between the acupuncture scores and the placebo scores. This

---

[35] Röschke, J., et al.; The Benefit from Whole Body Acupuncture in Major Depression, *Journal of Affective Disorders*, 57 (1): 73-81

was taken as evidence that acupuncture does not really work. However, if one reads the entire study, one learns that acupuncture and placebo *both* outperformed antidepressants!

Acupuncture in Combination with Other Methods

Acupuncture, although the most dramatic in appearance, is only one aspect of traditional Oriental medicine. It is usually used side-by-side with other modalities, such as herbal medicines, manual therapies, therapeutic exercise, etc. It would not be common to treat anxiety or depression, for example, with acupuncture alone. It has long been noted that a number of methods, acting in concert, bring the best results. However, a resourceful clinician can incorporate acupuncture into a treatment plan without necessarily becoming an expert in the whole scope of Asian medicine.

An interesting hybrid method was pioneered by Deborah Craydon and Warren Bellows[36] in which floral essences (covered in Chapter Ten of this book) are applied to acupuncture points, *creating a synergy of the two modalities*. Similarly, it is common in India for acupuncture needles to be dipped in homeopathic medicines before insertion, again blending the effects of two therapies.

Following are suggested acupuncture protocols for a variety of conditions. Needle acupuncture is of course not the only effective method. Non-needle stimulation of the points, whether by electrical stimulation, laser, magnetic fields, or mechanical pressure, has been found to be efficacious. This opens up the possibility of acupuncture points becoming part of a customized treatment plan even if the clinician is not a trained acupuncturist. Incorporating potentially useful points as part of the prescribed course of action can add another element that will increase chances of clinical success.

---

[36] Craydon, D., Bellows, W.; *Floral Acupuncture: Applying the Flower Essences of Dr. Bach to Acupuncture Points*, Crossing Press, 2005

In the point prescriptions, suggested stimulation is differentiated by (s) for sedation, (t) for tonification, and (n) for neutral. The nomenclature for the meridians is the abbreviation system currently in use throughout most of North America:

LU = Lung
LI = Large Intestine
ST = Stomach
SP = Spleen/Pancreas
HE = Heart
SI = Small Intestine
UB = Urinary Bladder
KD= Kidney
PC = Pericardium
TW = Triple Warmer
GB = Gall Bladder
LV = Liver

There may be more than one point formula recommended. In such cases, they are referred to as "Tx #1", "#Tx #2", etc. The #1 treatment plan should in most cases be tried first. If there is slow response (i.e., no improvement within three sessions), the next treatment formula should be implemented. Of course, the best results are gained by a carefully chosen set of points based on the review of the complete individual picture of the patient. The suggested courses of point treatment in this section are based on many of the most reliable point formulas listed in the "sacred formulas" as well as the more modern research of European leaders in the field.

Adrenal dysfunction contributing to symptoms:
- Tx#1: UB-62, SI-3 (n)
- Kd-7, UB-52, Sp-6, GV-6, CV-6/10/16, PC-7 (t)
- Tx#2: If pituitary is affected, CV-10/15/16/19, Kd-11/13, Sp-6, B-47/52, GB-5/37, UB-60
- Points for adrenals: Kd-7, UB-52, Sp-6, GV-6, CV-6/10/16, PC-7

Alzheimer's Disease:
- *Sea of Marrow* points: GV-1, GV-15, GV-16, GV-17 and GV-28
- TW-19, Kd-16 and Lv-7

Anxiety:
- Tx#1: PC-6/7, He-7, X-1, GV-21, Liv-3 (s) [Note: VGV-21 (classical GV-21) is MP Amygdaloid Nucleus—potent point for anxiety]
- Tx#2: SI-3, He-5, UB-62, Kd-6 (s)
- Obsessive thoughts: Sp-5, UB-8 (s)

Appetite, increased:
- Liv-3, Sp-6, He-7, St-13/41 (s)
- Ear: stomach, *shen men*, and upper lung point (for starch craving, especially)

Appetite, decreased:
- CV-12, Sp-4, St-45, PC-6 (t)

Attention Deficit Disorder:
- Tx #1: Ht-7, Ht-8, Lv-2, St-40, GB-34, PC-5 (s)
- Sp-6, St-36, Ht-7, Lv-3 (t)
- Tx#2: GV-20, PC-6, Ht-7, Lv-3/8, Sp-6 (n)
- Tx#3: Kd-3, UB-15/23

Attention Deficit Hyperactivity Disorder:
- Tx #1: Ht-7, Ht-8, Lv-2, St-40, GB-34, PC-5 (s)
- Sp-6, St-36, Ht-7, Lv-3 (t)
- Tx#2: GV-20, PC-6, Ht-7, Lv-3/8, Sp-6 (n)
- Tx#3: Kd-3, UB-15/23
- Tx#4: CV-13, Ht-7 (s)

Bipolar Affective Disorder:
- Tx#1: PC-7, LI-1, Sp-6, Kd-4, GV-15/20/23 (t)
- Tx#2: CV-4, CV-12, CV-22, GV-9
- Tx#3: PC-5, SI-3, St-40, Kd-1, GV-20
- Obsessive thoughts: Add Sp-5, UB-8 (s)

Cognition, poor:
- Tx#1: LI-4, Lv-3, Ht-7, PC-6, Sp-6, Kd-9, CV-14 (t)

Dementia, senile:
- *Sea of Marrow* points: GV-1, GV-15, GV-16, GV-17 and GV-28
- TW-19, Kd-16 and Lv-7

Depression:
- Tx#1: He-7, PC-6/7, X-1, X-8 (*An mian*) (s)
- Tx#2: Lu-1 (s) releases sadness
- Ear point: *shen men*
- Obsessive thoughts: Add Sp-5, UB-8 (s)

Hyperthyroidism contributing to symptoms:
- Tx#1: CV-22, CV-23, LI-4, X-2 (n)
- Tx#2: CV-13/19, GV-15/23, St-9/10, PC-6, Ht-7, LI-4, CV-6, TW-3 (s)

Hypothyroidism contributing to symptoms:
- Tx#1: TW-3, CV-22, LI-4, GV-15 (t)
- Tx#2: PC-6/8, UB-15/17/43, tip of C4 spinous process (t)

Insomnia:
- Tx#1: X-1, Ht-7, PC-6, Sp-6/9 (s); Sp-1, St-45(t)

Mental fatigue:
- Tx#1: LI-4, Lv-3, Ht-7, PC-6, Sp-6, Kd-9, CV-14 (t)

Obsession:
- Tx#1: Sp-5, UB-8 (s)

Night Terrors:
- Tx#1: X-1, Ht-7, PC-6 (s); Sp-1, St-45(t)

Premenstrual Syndrome:
- Tx#1: Sp-10, UB-67, Kd-2, CV-6 (moxa or t); Sp-6, Lv-2/3, Ht-7 (s)

Tourette's Syndrome:
- Tx#1: Ht-6/7/8; PC-6(s), GV-17/20 (n)

Bibliography

- Cho, Z., et al.; New Findings Of The Correlation Between Acupoints And Corresponding Brain Cortices Using Functional MRI, *Proc Natl Acad Sci USA* 1998 (95)
- Craydon, D., Bellows, W.; *Floral Acupuncture: Applying the Flower Essences of Dr. Bach to Acupuncture Points*, Crossing Press, 2005
- Hui, K.S., et al.; Characterization Of The "Deqi" Response In Acupuncture, *BMC Complementary and Alternative Medicine* 2007; (10):7:33
- Hui, K.S., et al.; Acupuncture Modulates The Limbic System And Subcortical Gray Structures Of The Human Brain: Evidence From fMRI Studies In Normal Subjects *Hum Brain Mapp* 2000; 9:13–25
- Hsieh J., et al.; Activation Of The Hypothalamus Characterizes The Acupuncture Stimulation At The Analgesic Point In Human: A Positron Emission Tomography Study, *Neurosci Lett* 2001; 307:105–8
- Röschke, J., et al.; The Benefit From Whole Body Acupuncture In Major Depression, *Journal of Affective Disorders* 2000; 57 (1): 73-81

7

MANUAL THERAPIES

It may seem ludicrous to include this modality in a book dealing with psychoemotional illness, inasmuch as manual therapies are associated mainly with musculoskeletal disorders. Of course, it could be argued that various massage and manipulative techniques could be of secondary value in relieving the muscle tension patterns associated with dysfunctional emotional states, but this is not the thrust of this chapter.

A number of systems of manual therapy have the potential to balance mental and emotional states as well as muscle groups. That they have been effective in this regard is corroborated by the many practitioners who have used them for years. We will first present the main systems and a brief description of their techniques. This will be followed by a checklist for assessing the core emotional problem to be reached with any type of therapy. Using body-based feedback to arrive at the most important sentiment, issue, trauma, or emotion is surprisingly valid and an efficient way to analyze a case. Moreover, locally treating an area of the body where a particular stress or locked emotion has settled or somatized *has the potential for creating a potent emotional healing response*. The emotional releases that people experience at the hands of body workers are common and powerful.

Overview of Manual Methods Used as Psychotherapy

Craniosacral Therapy

Cranial manipulation was formally developed by Dr. William Sutherland, D.O. in the early 1900's and came to be called cranial osteopathy. Sutherland noticed that the sutures of the skull seemed designed to allow for movement. He wondered why that would be if (as he had been taught in medical school) the cranial bones are completely fused in adults. Sutherland spent months researching the effects of jamming and immobilizing individual cranial bones. He used himself as a test subject and experienced a myriad of physical, mental and emotional symptoms as a result of self-induced cranial restrictions. He then spent years developing and refining techniques to correct the conditions he had discovered. Cranial osteopathy focuses on restoring and maintaining osseous mobility. The concept of cranial bone movement is still controversial in the United States.

Craniosacral therapy (CST) was developed in the 1970's by Dr. John Upledger, D.O. Dr. Upledger's most significant contribution to cranial manipulation may have been to shift the focus of therapy from the bones themselves to the connective tissues of the body, particularly the meninges. CST uses bones as handles to gently release restrictions in the soft tissues of the craniosacral system. Restoring elasticity and movement in the membranes that surround the brain and spinal cord can bring about dramatic effects in the central nervous system. Since all connective tissue in the body is contiguous, these same techniques can be used to release restrictions and adhesions anywhere in the body.

The work has proven to be of benefit to patients with mental and emotional issues as well as physical ones. In a study with 22 Vietnam veterans suffering from posttraumatic stress disorder (PTSD), CST was successful in improving or eliminating symptoms in all test subjects.

How does a manual technique affect PTSD? Dr. Upledger suggests that focusing on the Reticular Activating System of the brain during CST sessions can calm the nervous system and help eliminate PTSD symptoms such as hypervigilance and hyper-responsiveness. Improving circulation of blood and cerebro-spinal fluid by releasing adhesions and restoring normal movement patterns increases available nutrition to the brain and helps remove toxic waste products. This improves functionality in both hemispheres and may help to eliminate flashbacks. Relieving tension and immobility in articulations of the head and neck improves circulation, increases normal range of motion and seems to alleviate insomnia. CST practitioners have identified three specific articulations that are often found to be abnormally compressed in individuals suffering from endogenous depression. This "Triad of Depression" includes the spheno-basilar articulation, the atlanto-occipital articulation and the lumbo-sacral articulation. All three are addressed in a basic craniosacral treatment. Although the focus is on manual manipulation, it should be noted that craniosacral treatment can also include dialogue.

Several possible mechanisms exist, but the most powerful component of the craniosacral approach may be its gentleness. The amount of pressure most commonly used in treatment is about five grams (the weight of a nickel). This light-touch treatment tends to produce profound relaxation and altered awareness, which can lead to valuable insight. Emotional catharsis and insight and are also reported effects of other light-touch methods such as polarity therapy. Gentle contact, with the intention of supporting rather than directing the process, may allow CST practitioners to bypass the body's normal defense mechanism. This gives access to deeper tissues (and deeper "issues") that more aggressive methods are unlikely to affect. The combination of an empowering, client-centered approach and gentle, myofascial manipulations may be a major key to CST's effectiveness.

Craniosacral therapy therapy is used by osteopaths, massage therapists, naturopaths, chiropractors, physical therapists, occupational therapists and many others. The Upledger Institute, which represents only one of several versions of craniosacral therapy, presents about 500 classes each year throughout the world.

Applied Kinesiology (AK) is a chiropractic assessment technique that uses manual resistive muscle strength testing to identify a wide variety of health problems. It is based on the concept that weakness in certain muscles may correspond to specific diseases, dysfunctions, structural imbalance, energy blockage, allergies or nutritional deficiencies. AK was developed in 1964 by Dr. George Goodheart Jr. (1918 - 2008), a Michigan chiropractor who noticed that postural distortions are often associated with weak muscles. He suggested that interventions could be identified and tested using AK muscle testing. The most effective treatment could then be chosen based on its ability to make muscles stronger and improve postural distortions.

AK practitioners (mainly doctors) first learn to recognize a muscle that is weak or "unlocked" when tested. The person being tested holds a limb in a specified position while the "tester" exerts mild pressure against the limb. The muscle being tested should be able to hold that shortened position with little effort. If the muscle moves, trembles or shows signs of weakness, the practitioner works to determine the source of the weakness and the optimum treatment to correct it.

For each muscle studied, practitioners learn the corresponding acupuncture meridian as well as nutritional influences and various problems or medical conditions associated with weakness in that particular muscle.

They are then taught how to correct the muscle weakness and simultaneously address the related condition through nutrition, chiropractic adjustment, acupuncture stimulation, exercise, soft tissue manipulation or neurological reflex techniques.

Some practitioners use glass vials containing a food, supplement, medicine, allergen or toxin (or the energetic signature of some chosen test substance) in order to evaluate the muscle response. If the vial contains water that has been imprinted with an energetic signature, as in a homeopathic preparation, the body responds to it as if it were the actual substance. The person being tested holds the vial while an indicator muscle is tested. A weak muscle response indicates the substance is weakening to the subject's body and may be the cause of his or her problem. When a previously weak muscle tests stronger in the presence of the vial, the substance may be a useful part of the subject's treatment plan. The muscle test is seen as a way to non-verbally communicate with the subject's autonomic nervous system and/or unconscious mind.

In 1976 the *International College of Applied Kinesiology* was founded to promote research and teaching in the area of applied kinesiology muscle testing. Currently, AK is practiced by chiropractors, massage therapists, physical therapists, naturopaths, dentists, nutritionists, nurse practitioners, medical doctors, and other providers.

Touch for Health Kinesiology (TFHK) was created by John F. Thie, DC. After completing Goodheart's applied kinesiology course and experiencing the effectiveness of his techniques, Dr. Thie saw value in teaching the method to non-professionals. He borrowed from the AK material to form a streamlined, easy to learn version called "Touch For Health" (now called "Touch For Health Kinesiology"). With this new format, individuals with no medical training could take a weekend seminar and be able to use the material immediately to help themselves and their families. The first TFH Manual was published in 1973, and it has been adopted by thousands of chiropractors, acupuncturists, massage therapists, and laypeople. The *Touch For Health Kinesiology Association* was formed to promote and support this work.

Like Applied Kinesiology, TFHK is based partly on acupuncture/acupressure principles. Each of the muscles studied is associated with one of 14 acupuncture meridians. From a TFHK perspective, a problem in muscle function often reflects an excess or deficiency in that muscle's associated meridian. The root cause of the meridian/muscle imbalance may be energetic, emotional, nutritional or structural.

Students then learn a variety of ways to strengthen, correct or "switch on" an unlocked muscle. Techniques which require a medical or chiropractic license were eliminated from the system and replaced with safe and effective alternatives. These include soft tissue manipulation, light contact on specific acupuncture points, tracing the associated meridian, massaging neurolymphatic (Chapman's) reflexes, lightly holding neurovascular (Bennett's) reflexes, simple exercises and more.

By applying the appropriate techniques to key areas in order to restore balance, the muscle strength quickly returns on re-examination. Normalizing the person's energetic system tends to create a more normalized functioning of the musculoskeletal system and even organic function in general. However, the balancing effect is often most profound at the mental-emotional level, and this is where TFHK is relevant in our discussion.

There are many "off-shoots" and variations of AK and TFHK that focus on specific areas of interest, medical conditions or treatment approaches. The common denominator in these "specialized kinesiologies" is reliance upon manual muscle testing, which has also been referred to as "energy testing" and "brain response testing". Following are some examples.

Edukinesthesia is a type of AK that is used to detect the cause of learning difficulties and poor concentration. It has applications for emotional as well as cognitive problems. Popular variations of this method include "Educational Kinesiology", "Behavioral Kinesiology" and "One Brain". These methods are widely used to help children and adults with dyslexia, ADHD and related difficulties.

Contact Reflex Analysis® (CRA) uses manual resistive muscle testing to evaluate the energetic status of the body. The practitioner performs a muscle test (usually using the deltoid) while touching one or more reflex points on the client's body. Each reflex point corresponds to an organ, structure or system. How the muscle reacts to that point reflects the energetic health of the associated organ, structure or system. The weaker the muscle, the greater degree of imbalance is indicated. This type of assessment is similar to "therapy localization" in applied kinesiology and "circuit locating" in Touch for Health Kinesiology.

Jaffe-Mellor Technique (JMT™)
Practitioners of MRT procedures developed by Jaffe and Mellor claim to be distinct from all other styles of Applied Kinesiology because of a) blind testing to validate the accuracy of the test, b) a high level of specificity in the kind of question it can present to the body of the patient, and c) the ability to frame very specific corrective instructions to the nervous system of the patient.

JMT™ practitioners encourage the patient to relax and refrain from conversation during treatment. The emphasis is on non-verbal communication with the subject's autonomic nervous system. Part of this communication involves the nervous system being stimulated by tapping or electrical stimulation along the paraspinal muscles. Reactions are then evaluated and interpreted by means of manual resistive muscle testing. This testing can guide the practitioner to find very specific areas of internal dysfunction as well as specific treatment solutions.

Emotional Stress Release
Emotional Stress Release (ESR) is a simple Touch For Health protocol that can be taught to the patient for self-application. It involves lightly touching neurovascular reflex points (the frontal eminences of the forehead) and seems to have a calming effect on the forebrain, where solutions to problems are processed. By utilizing this neuromuscular connection while thinking about a particular emotional issue, the process of forming new reactions to the stressful situations is made easier.

This may be the conscious application of an instinctive process. What is more natural than placing your palm on your forehead when under stress?

Accurate Muscle Testing

To use ESR more effectively, follow these basic guidelines for accurate muscle testing:

1. Use a maximum of two pounds of pressure when muscle testing. In general, lighter pressure is better. We are assessing the ability of the muscle to "lock" (hold a limb in place without undo effort) not the strength of the muscle. Only a slight pressure is needed.

2. Do not exert pressure for more than two seconds. With extended resistance, even an optimally functioning muscle will fatigue and unlock.

3. If the limb gives way when tested, it is not necessary to move it more than two inches. A muscle either locks or it doesn't.

4. Prepare the person being tested by demonstration the direction you intend to push and giving ample warning. For example, "Ready...and...hold" is effective instruction. The intent is to work with the person being tested and to avoid any surprise.

Finding an Indicator Muscle

It is also important to find a suitable indicator muscle. An indicator muscle can be any properly functioning muscle that is convenient to use. Evidence of proper functioning would be a good strong lock in the initial test. The muscle also needs to be able to unlock when appropriate. Manually stimulating the muscle by repeatedly squeezing or pinching fibers near the center of the belly should unlock the muscle by activating the spindle cell mechanism. Asking the person being tested to think of an emotionally stressful event, or French fries cooked in rancid oil, should similarly unlock the muscle. If not, choose a different indicator muscle.

Five Element Emotions

Once an indicator muscle is chosen, the *five elements* concept can be used to identify an active but possibly unconscious emotional stressor. The technique is quite simple.

The tester verbalizes the names of the five elements from Chinese medicine, one at a time. After each element is named, the indicator muscle is tested. If there is stress involving a particular element, the muscle will unlock when that element is named. It is not necessary for the person being tested to have any conscious knowledge about the five elements.

Each element is associated with two or more meridians (energy channels in acupuncture theory). Once the element is identified (by a muscle unlocking), the tester verbalizes the names of the associated meridians, testing the indicator after each one. The indicator will unlock when the involved meridian is named.

Each meridian has certain emotions associated with it. The TFHK list of emotions is minimal; however, more advanced kinesiology courses offer more extensive lists of five-element emotions. With the relevant meridian identified, the tester simple reads through the list of emotions associated with that meridian, testing the indicator after each emotion. If the person being tested reacts to a particular emotion, the indicator will unlock when the emotion is verbalized.

Now that the emotion is identified, ESR can be used to resolve the stress. The person just tested can hold the name of the emotion in his mind while either he or the tester lightly touches the subject's frontal eminences. Sometimes the relevant situation comes to mind immediately, but that is not necessary. When a noticeable shift in mood occurs, or when verbalizing the emotion no longer causes the indicator muscle to unlock, the process is complete.

Below is a list of five element emotions presented in a course called Professional Kinesiology Practitioner (PKP) training, which was developed by Dr. Bruce Dewe.

Five Element Emotion/Meridian Chart
Central and Governing Meridians

Meridian	Expression	Affliction
Central	Success	Overwhelmed
	Self respect	Shame, Shyness
Governing	Supported Trust Honesty	Unsupported Distrust Dishonesty, Embarrassment

Element: Fire

Polar Expressions: Joy, Love/Hate, Hysteria/Tranquility, Excitability/Calm

Meridian	Expression	Affliction
Heart	Love Forgiveness Security	Anger Insecurity
Small Intestine	Joy Appreciated	Sorrow Unappreciated
Pericardium	Relaxation Satisfaction Renunciation of past	Tension Stubbornness Worried Regret, Remorse
Triple Warmer	Lightness Serving Hope Elation Buoyancy	Heaviness Depression Humiliation Hopelessness Despondency Grief Loneliness Solitude

Element: Earth
Polar Expressions: Sympathy/Empathy, Worry,
Doubt/Sympathy, Criticism, Envy/Empathy

Meridian	Expression	Affliction
Spleen	Faith in the future Approved Security	Anxiety about the future Disapproved Rejection
Stomach	Content Reliable Tranquility Empathy Hunger	Disappointment Unreliable Bitterness Greed Sympathy Emptiness Deprivation Disgust Nausea

Element: Metal
Polar Expressions: Guilt, Grief, Regret, Depression/Enthusiasm,
Sadness/Cheerfulness

Meridian	Expression	Affliction
Lung	Tolerance Cheerful Humility Modesty	Intolerance Depressed False Pride Haughtiness Scorn Prejudice Disdain Contempt
Large Intestine	Self Worth Mercy	Guilt Unmerciful

Element: Water
Polar Expressions: Fear, Anxiety/Courage, Loss of self-confidence/Confidence

Meridian	Expression	Affliction
Kidney	Sexual security Creative security Loyalty Decisiveness	Sexual indecision Creative insecurity Disloyal Indecisiveness
Bladder	Peace Harmony	Restlessness Fear Impatience Frustration Anxiety Terror Dread

Element: Wood
Polar Expressions: Anger/Happiness; Resentment/Contentment; Impatience/Patience; Frustration/Forbearance

Meridian	Expression	Affliction
Liver	Happiness Contentment Cheer	Unhappiness Distress Righteous indigna-tion
Gall Bladder	Reaching out with love Humility Adoration	Anger, Rage, Fury Pride Wrath

Bibliography:

- Frost, R., *Applied Kinesiology: A Training Manual and Reference Book of Basic Principles and Practices*, North Atlantic Books 2002
- Thie, J.; *Touch For Health: The Complete Edition*, DeVorss & Company 2005
- Upledger, J., *Your Inner Physician and You,* North Atlantic Books, 2nd Edition1997
- Upledger, J., *Craniosacral Therapy: Touchstone for Natural Healing,* Georg Thieme Verlag, Germany 2002
- Versendaal, D.A.; *Contact Reflex Assessment and Applied Trophology*, D.A. Versendaal, Holland (MI) 1990

8

BODY-CENTERED PSYCHOLOGY

The psycho-emotional-neuro-endocrine (PENE) mechanism, as the reader now sees, is the system to be adjusted for optimal effects in treatment. Now that some basic concepts regarding the body-mind connection have been illustrated in Chapter Seven, we can look at some systems of psychotherapy that have borrowed from the manual therapies in much the same way that manual therapies such as Touch For Health Kinesiology have drifted into the realm of psychotherapy.

Neuro-linguistic Programming (NLP) was promoted in the 1970's as a fast and effective form of psychotherapy. Co-founders Richard Bandler and John Grinder, a linguist, saw relationships between neurological processes, thought patterns, language, and learned behavior. They believed that better understanding of these relationships, along with training in self-awareness and communication, could help people to re-program limiting or self-defeating patterns of thought and behavior. They were largely ignored by conventional psychology because of a lack of empirical evidence.

NLP was later promoted as a "pathway to excellence". Practitioners and teachers focused on the concept of "modeling" as a powerful and accessible mode of learning. They studied behaviors of very successful people in order to see how those people achieved their results. That information was then organized according to NLP concepts and perspectives. This provided a foundation for anyone to learn how to mimic those successful behaviors and achieve similar results. NLP offered a way not only to "program out" limiting behavior patterns but to replace them with more functional alternatives, which could help people to reach their goals and achieve their full potential.

The founders of NLP were strongly influenced by psychiatrist and psychologist Milton Erickson and his unique understanding of hypnosis. NLP continues to be popular with hypnotists, self-help enthusiasts, life coaches, and in the fields of organizational development and management training.

Eye Movement Desensitization and Reprocessing (EMDR) is a powerful but controversial form of psychotherapy. It is eclectic in the sense that it integrates components of several conventional therapies including psychodynamic, cognitive-behavioral, interpersonal, and body-centered therapy.

The initial objectives are to develop goals and directions, as well as to determine whether EMDR is the most appropriate treatment for that particular client. Stability and coping skills are also assessed. If they are found to be inadequate, that issue is addressed in therapy until the necessary skills are in place.

Once the client is ready to continue, he or she identifies one or more sources of anxiety or discomfort. Several sessions are then devoted to targeting those beliefs, emotions, memories, or sensations that are causing the discomfort. The client is also asked to identify a positive belief, image, statement or idea.

One essential concept is the use of "dual stimulation" which involves eye movements (usually side to side movements on a horizontal plane following the therapist's finger), tones or tapping. The client is asked to focus on a past memory, present sensation or anticipated future event while also attending to the external stimulation. This is the "re-processing" phase, and the client is likely to experience noticeable shifts in his or her personal experience. Insights may emerge, or the emotional charge around an issue may dissipate. Physical sensations such as pain or tension often change or disappear.

When the negative sensation or emotion is no longer noticeable, the dual stimulation is applied in a similar way to the positive statement, image or goal. The client usually reports having more confidence in the positive goal after treatment. Several issues, memories, sensations or beliefs can be addressed in a single session.

Later sessions may re-evaluate previous work and address future situations for which the client feels unprepared. The complete treatment process clears anxiety and distress from past, present and future scenarios. Clients report more confidence as well as increased awareness that often generates immediate behavioral changes. EMDR is effective as a relatively short-term therapy. A single traumatic event may be cleared in one to three sessions. On-going trauma or abuse can take longer.

Thought Field Therapy (TFT) is primarily used to treat emotional trauma and posttraumatic stress disorder (PTSD). Roger Callahan, the American psychologist who developed TFT in 1981, claims that it can relieve a wide variety of physical and psychological complaints including anxiety, depression, alcohol abuse, anger, bereavement, chronic pain and even serious heart problems.

Callahan posits that each emotional problem is associated with what he calls a "thought field". Precisely encoded bits of information (called *perturbations*) in this field are activated whenever a person tunes into the thought field by thinking about the associated problem or situation. In TFT, perturbations are considered to be the root cause of all negative emotions.

Each perturbation also corresponds to a particular meridian point on the body. The TFT practitioner taps these meridian points in a precise sequence or algorithm (specific to the problem) in order to unblock energy flows, eliminate perturbations and relieve the emotional upset. Whatever the mechanism, TFT seems to have benefit for physical problems as well as emotional ones.

The dearth of scientific evidence on the effectiveness of Thought Field Therapy has led the American Psychological Association to state that TFT lacks a scientific basis. Practitioners say that the positive results they get with clients are more important than double blind studies.[37]

Timeline Repatterning® was developed by Dr. George Roth. It is a process of guided visualization in which patients re-imagine past traumas and negative experiences from a state of higher understanding and positive intention. The patient is helped to "re-write" his or her own story and to make the new, more positive, version feel as real as possible.

With this new perspective, the patient can re-frame memories, change old beliefs and alter patterns of behavior. Timeline Repatterning empowers patients to recognize their ability to choose any belief, any time, about any part of their lives. Rather than focusing on the negative, it lays a foundation for new possibilities to be imagined and experienced.

Today, this basic idea is often integrated into other approaches. Imagine utilizing NLP techniques to enhance this Repatterning process. One could easily add the dual stimulation of EMDR to access the autonomic nervous system or tapping from TFT to more directly involve the meridian system. With a vast array of powerful techniques available, there are many eclectic practitioners in the emerging field of energy psychology.

Emotional Freedom Technique (EFT) was developed by Gary Craig, who is neither a psychologist nor a licensed therapist. EFT can be considered either an offshoot or an evolution of Thought Field Therapy (TFT). Both approaches have continued to grow and develop since the birth of EFT in the mid 1990's.

EFT and TFT share the basic philosophy that all emotional and psychological problems stem from disruption in the body's en-

---

[37] To explore the current research on this and similar energy based therapies, consult the Association for Comprehensive Energy Psychology at http://www.energypsych.org.

ergy system. The primary difference is that TFT is very emphatic about tapping a specific set of points in a specific order for a particular type of problem. TFT practitioners are required to learn ten to fifteen different protocols. Gary Craig, however, proposed that all of the meridians could be addressed in a single protocol. EFT uses one set of meridian points for virtually any problem or condition. The order in which the points are addressed is not considered important.

EFT has been embraced by a wide variety of health professionals and laypersons. Its simplicity and effectiveness as a self-help method earns it further description here. At this writing, the complete basic EFT manual is available as a free download from Gary Craig's website, www.emofree.com.

Using EFT

Before beginning any new process, self-assessment can be helpful. Rating one's physical, mental and emotional state on a scale of one to ten provides a useful reference for later comparison. If there is a problem or complaint, any discomfort should also be rated on a scale of one to ten.

The EFT protocol is called "the basic recipe". The first two parts of the recipe, which should be sufficient for most problems, are described below.

1. The set-up is a simple technique to correct what is called a "psychological reversal" (a term coined by Dr. Roger Callahan). Psychological reversals are caused by negative thinking, conscious or unconscious, and Craig estimates that this condition is present 40% of the time. Psychological reversals represent a serious dysfunction in the body's energy system. If the condition is present, Craig claims that it blocks the effectiveness of EFT *or any other healing method*. His solution is to routinely include the correction as part of the treatment.

The set-up consists of an affirmation that is repeated three times while simultaneously stimulating a reflex area on the chest,

101

which he calls the "sore spot". One side or both sides can be stimulated, preferably by digital massage. The reflex area can be found between the second and third rib and about halfway between the sternum and glenohumeral articulation. The area will often be tender. An alternative to massaging the sore spot is tapping a point on the center of the ulnar border of the hand. Note the similarity to the "dual stimulation" concept of EMDR.

The affirmation always includes a statement about loving and accepting one's self. The prelude to that statement varies with the presenting complaint. For the purpose of illustrating this process, let's decide that our imaginary client suffers from a *fear of relationships*.

The patient or client would be asked to repeat:

"Even though I have this *fear of relationships*, I totally and completely love and accept myself."

The client would be asked to repeat it three times while massaging the sore spot and may be encouraged to vocalize with more enthusiasm or determination. He or she would then proceed to the sequence.

2. The sequence is a set of acupuncture points, which are tapped one at a time, for about two to five seconds each, at a rate of about four taps per second. As each point is stimulated, the client verbalizes a simple reminder phrase such as, "*this fear of relationships*". Tapping may be performed on one side of the body or both sides of the body simultaneously.

- The first point is at the medial end of the eyebrow, where the nose meets the eyebrow.
- The second point is level with the eye on the lateral side of the orbit.
- The third point is directly underneath the eye at the infra-orbital foramina.
- The fourth is at the center of the upper lip.

- The fifth point is at the center of the lower lip. Both lips can be tapped simultaneously.
- The sixth point is where the clavicle meets the sternum, under the clavicle.
- The seventh point is just under the breast, about one inch below the nipple in men.
- The eighth point is on the side of the trunk, level with the nipple on men.

After tapping each of the points while repeating the reminder phrase, the client will pause, take a deep breath and re-evaluate the presenting complaint. If the discomfort is reduced but not eliminated, the client will repeat the sequence using a reminder phrase such as, *"remaining fear"*. In a great many cases, the discomfort will be gone.

Craig encourages his students and clients to try EFT with any problem that may arise. The technique is very safe, and the results can be surprising. For physical problems, the same procedure would be used. A set-up affirmation might be, "Even though I have this *back pain*, I totally and completely love and accept myself."

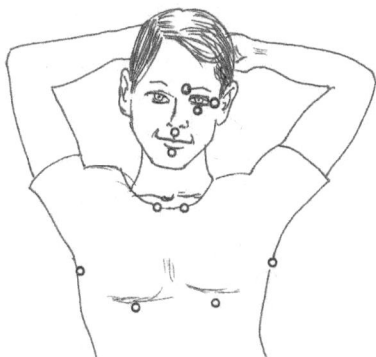

There are additional techniques for dealing with more persistent issues.

Interested clinicians are encouraged to explore the abundance of educational resources and training available today.

Bibliography:

- Connolly, S.; *Thought Field Therapy: Clinical Applications, Integrating TFT in Psychotherapy,* George Tyrrell Press 2004
- Craig, G.; *The EFT Manual,* Energy Psychology Press 2008
- Feinstein, D.; Energy Psychology: A Review of the Preliminary Evidence, *Psychotherapy: Theory, Research, Practice, Training* 2008; 45(2): 199-213
- Kavanagh, D. J., Freese, S., Andrade, J., & May, J. "Effects of visuospatial tasks on desensitization to emotive memories", *British Journal of Clinical Psychology* 2001; 40(3), 267-280
- Lee, C.W., G. Taylor, and P. Drummond, The active ingredient in EMDR; is it traditional exposure or dual focus of attention? Clinical Psychology & Psychotherapy, 2006. 13: p. 97-107
- Oschman, J.; *Energy Medicine; The Scientific Basis,* 1st Edition, 2000, Churchill Livingstone
- Pert, C.; *Molecules of Emotion,* 1999 Simon & Shuster, New York
- Ruden, R.; A Model for Disrupting an Encoded Traumatic Memory, *Traumatology* 2007; 13:71-75
- Scaer, R., *The Body Bears the Burden,* 2001 Haworth Medical Press, New York
- Shapiro, F.; *Eye Movement Desensitization and Reprocessing (EMDR): Basic Principles, Protocols, and Procedures,* 2nd Edition, 2001 The Guilford Press
- Shapiro, F. & Maxfield, L. (2002). "Eye Movement Desensitization and Reprocessing (EMDR): Information Processing in the Treatment of Trauma", *Journal of Clinical Psychology,* 58, 933-948.
- Smyth, N.J. and A.D. Poole, "EMDR and cognitive-behavior therapy: Exploring convergence and divergence, in EMDR as an integrative psychotherapy approach: Experts of diverse orientations explore the paradigm prism", F. Shapiro, Editor. 2002, American Psychological Association: Washington, DC. p. 151-180.

9

OLFACTORY THERAPY

Aromatherapy is a generic term for the use of the volatile oils of plants to treat or prevent disease and enhance health, particularly improving cognition and mood. The aromatherapist applies these oils in three ways: *aerial diffusion* to spread the fragrance through the immediate environment; *direct inhalation* by the patient; and *topically* in baths, compresses, and rubbed into the skin as part of a massage.

An illustration of the latter: Even young children can be treated safely with these volatile oils if used topically, and with care. A commonly used compound for hyperactive children is one drop each of the following essential oils, mixed in 1/8 cup olive oil:
- Chamomile
- Sage
- Rosemary
- Lavender

Applied to the soles of the feet and the spine at bedtime, this has a calming effect in many individuals who have not even responded well to Ritalin and other drugs.

The term *psychoaromatherapy* has entered the medical dictionary, with a meaning of "the use of fragrances to manipulate mood, with the goal of bringing about an enhanced sense of well-being". We are referring to it here as *Olfactory Therapy* since it is the olfactory sense that provides the strongest communication with the brain to allow the effects of the oils to take action on the mind of the subject.

The essential oils that concern us for use in the most common mental/emotional symptoms are these:

- African Lily of the Valley
- Angelica
- Arabian Wild Rose
- Basil
- Chamomile
- Clary sage
- Egyptian Musk
- French Lavender
- Lavender
- Mysori Sandalwood
- Rose
- Rosemary
- Sandalwood
- Tunisian Amber
- Tunisian Frankincense
- Tunisian Honeysuckle
- Tunisian Jasmine
- Tunisian Myrrh
- Tunisian Patchouli
- White Opium
- Ylang Ylang

Following is a graph for determining which oil contains the needed actions for the case at hand.

Essential Oil	Anti-anxiety	Anti-depressant	Relieves Fatigue	Improves Cognition	Mood Elevator	Sedative	Tonic	Insomnia	Anger	Releasing Negativity	Headaches
African Lily of the Valley					■				■	■	■
Angelica	■	■				■					
Arabian Wild Rose			■				■			■	
Basil	■	■	■				■				
Chamomile	■					■		■			
Clary sage	■	■						■			
Egyptian Musk			■	■			■				
French Lavender					■					■	
Lavender	■	■						■			■
Mysori Sandalwood	■	■				■					
Rose						■		■			

Essential Oil	Anti-anxiety	Anti-depressant	Relieves Fatigue	Improves Cognition	Mood Elevator	Sedative	Tonic	Insomnia	Anger	Releasing Negativity	Headache
Rosemary		■	■	■			■				
Sandalwood		■						■			
Tunisian Amber			■	■			■			■	
Tunisian Frankincense	■			■			■				
Tunisian Honeysuckle					■						
Tunisian Jasmine	■	■	■	■	■		■	■		■	
Tunisian Myrrh		■		■	■					■	
Tunisian Patchouli			■								
White Opium							■				
Ylang Ylang	■	■				■		■			

Bibliography

- Chen, SW., Min, L., Li, WJ, Kong, WX, Li, JF, Zhang, YJ; "The Effects of Angelica Essential Oil in Three Murine Tests of Anxiety". *Pharmacology, Biochemistry, and Behavior* 79 (2): 377–82.
- de Almeida, R., Motta, SC, de Brito, F., Catallani, B, Leite J.; "Anxiolytic-Like Effects of Rose Oil Inhalation on the Elevated Plus-Maze Test in Rats". *Pharmacology, Biochemistry, and Behavior* 77 (2): 361–4.
- Lis-Balchin M., Hart S.; "Studies on The Mode of Action of the Essential Oil of Lavender (Lavandula angustifolia)". *Phytotherapy Research* 13 (6): 540–2.
- Pisseri, F., Bertoli, A., Pistelli, L.; "Essential Oils in Medicine: Principles of Therapy". *Parassitologia* 50 (1-2): 89–91.
- Price, L., Price, S.; *Aromatherapy for Health Professionals*, Churchill Livingstone, 3rd Edition 2006
- Shaw D., Annett J., Doherty, B., Leslie, J.; "Anxiolytic Effects of Lavender Oil Inhalation on Open-Field Behaviour in Rats". *Phytomedicine* 14 (9): 613–20.

10

FLOWER ESSENCES

Edward Bach, M.D. was a bacteriologist in England who made important contributions to homeopathic medicine with seven "bowel nosodes" that he researched. These are medicines made from the flora found in the colon under different conditions, and which are associated with various systemic illnesses. Bach (pronounced "bash") found that these samples, when processed in the typical manner of making homeopathic medicines, acted as corrective agents in illnesses that were characterized by changed conditions within the intestinal tract or were caused by the microorganisms growing there.

The second half of his career, though, was in a different direction. He retired to the countryside and spent the rest of his life experimenting with flowers and their effects on the psyche. Rather than using the roots, stems, and leaves of plants as in most botanical medicines, he observed the properties of the flowering blossoms. Instead of the dilution/succussion method of preparing homeopathic medicines (see Chapter Thirteen), he developed a way of simply suspending the flowering tops in pure water and sunlight for a day. Dr. Bach was able to match up thirty-eight flower essences with their resonant symptoms and characteristics in the patients who needed them.

Since that time, the Bach Flower Essences have been used successfully by countless individuals to create positive changes in unwanted personality characteristics. They are often likened to homeopathic medicines. Their use, while not technically homeopathic in nature, does involve small doses, and they are prescribed on an individual basis; some homeopaths also prescribe Bach flowers.

Dr. Bach saw mental or emotional illness as a conflict between the personality and what he saw as the "higher self" of the individual. Therapy is successful only when the higher self is restored and the psyche is balanced, a point of view shared by enlightened conventional psychotherapy as well. However, rather than rely solely on creating new insights through therapy, or conditioning against emotional trauma, Bach's method strives to neutralize very specific conditions of the psyche as they are expressed: self-centeredness, fear of losing control, jealousy, etc.

Up to six flower essences can be combined in order to make a more perfect "fit" for the person. Following is a list of the major characteristics associated with each flower essence:

Brief Indications for the 38 Bach Flower Remedies

AGRIMONY: Mental torture, worry that is concealed from others ("smiling depressive anxiety")
ASPEN: Vague fears of unknown origin; anxiety; apprehension
BEECH: Intolerance; criticism, passing judgments
CENTAURY: Weak willed; too easily influenced
CERATO: Distrust of self; doubt of one's ability; foolishness
CHERRY PLUM: Desperation; fear of losing control of the mind: dread of doing some frightful thing
CHESTNUT BUD: Failure to learn by experience; lack of observation in the lessons of life, hence the need of repetition
CHICORY: Possessiveness, self-love; self-pity
CLEMATIS: Indifference; dreaminess: inattention
CRAB APPLE: The cleansing essence; despondency; despair
ELM: Occasional feeling of inadequacy; despondency; exhaustion from over-striving for perfection
GENTIAN: Doubt; depression; discouragement
GORSE: Hopelessness; despair
HEATHER: Self- centeredness; self-condemning
HOLLY: Hatred; envy; jealousy; suspicion

HONEYSUCKLE: Dwelling upon thoughts of the past; nostalgia; homesickness

HORNBEAM: Tiredness; weariness; mental and physical exhaustion

IMPATIENS: Impatience, irritability; extreme mental tension

LARCH: Lack of confidence; anticipation of failure; despondency

MIMULUS: Fear or anxiety of a known origin

MUSTARD: Black depression, melancholia; gloom

OAK: Despondency; despair; but never-ceasing effort

OLIVE: Complete exhaustion; mental fatigue

PINE: Self-reproach; guilt feelings; despondency

RED CHESTNUT: Excessive fear; anxiety for others

ROCK ROSE: Terror; extreme fright

ROCK WATER: Self-repression; self-denial; self-martyrdom

SCLERANTHUS: Uncertainty; indecision; hesitancy; unbalanced

STAR OF BETHLEHEM: Aftereffect of shock, mental or physical

SWEET CHESTNUT: Extreme mental anguish; hopelessness. despair

VERVAIN: Strain; stress; tension; over-exhaustion

VINE: Dominating; inflexible; ambitions

WALNUT: Oversensitive to ideas and influences; the link-breaker

WATER VIOLET: Pride; aloofness

WHITE CHESTNUT: Persistent unwanted thoughts; mental arguments and conversations

WILD OAT: Uncertainty; despondency; dissatisfaction

WILD ROSE: Resignation; apathy

WILLOW: Resentment; bitterness

RESCUE REMEDY: The "first aid" or emergency essences of Clematis, Cherry Plum, Impatiens, Rock Rose, and Star of Bethlehem in combination.

Using flower essences is essentially prescriptive. The clinician simply matches up the most indicated remedies with the predominant characteristics or symptoms. An expedient to this is having the patient or client fill out a questionnaire. The flower essences indicated for that person will be revealed by having the greatest scores.

Questionnaire To Determine Bach Flower Essences

Rate each question in the flowing manner:
0 = Not at all
1 = Sometimes
2 = Often
3 = Very much

Agrimony
____ Do you hide your worries behind a cheerful, smiling face to conceal your pain from others?
____ Are you distressed by arguments and quarrels, often "giving in" to avoid conflict?
____ When you feel life's pressures weighing you down, do you often turn to food, work, alcohol, drugs or other outside influences to help you cope?

Aspen
____ Do you have feelings of apprehension or anxiety without knowing why?
____ Do you feel that something bad may happen but you are not sure what?
____ Do you wake up with a sense of anxiety of what the day will bring?

Beech
____ Are you annoyed by the habits and shortcomings of others?
____ Do you find yourself being overly critical and intolerant, usually looking for what someone has done wrong and not right?
____ Do you prefer to work or be alone as the seeming foolishness of others irritates you?

Centaury

_____ Do you often neglect your own needs in order to please others?

_____ Is it difficult to say no to those who impose upon your good nature?

_____ Do you tend to be easily influenced by those stronger in nature than yourself?

Cerato

_____ Do you constantly second-guess your own decisions and judgment?

_____ Do you often seek advice and confirmation from other people, mistrusting your own intuition?

_____ Do you change direction often, even after asking advice, because you feel confused or unsure?

Cherry Plum

_____ Are you afraid you might lose control of yourself mentally, emotionally or physically?

_____ Do you fear that you may think or do something that you feel is wrong?

_____ Do you fear you may hurt yourself or others or become violent and explosive?

Chestnut Bud

_____ Do you find yourself making the same mistakes over and over again such as choosing the wrong type of partner or staying in a job you dislike?

_____ Do you fail to learn from the mistakes or experience of others?

_____ Do you wish you would not repeat the same patterns again and again?

Chicory

_____ Do you need to be needed and want your loved ones to be close by?

_____ Do you feel unloved and unappreciated by your loved ones?

_____ Are you possessive of those you care for, feeling you know what is best for them?

Clematis

____ Do you often feel spacey and absent minded?

____ Do you find yourself preoccupied and dreamy, unable to concentrate?

____ Are you drowsy and listless, sleeping more often than necessary?

Crab Apple

____ Are you obsessed with cleanliness or feel toxic or contaminated?

____ Are you embarrassed and ashamed of yourself or feel physically unattractive?

____ Do you tend to concentrate on small physical conditions such as pimples or marks?

Elm

____ Do you feel overwhelmed by your responsibilities?

____ Do you feel it is too difficult to handle all the many tasks ahead of you?

____ Do you become depressed and exhausted when faced with your everyday commitments?

Gentian

____ Do you become discouraged and depressed when things go wrong?

____ Are you easily disheartened when faced with difficult situations?

____ Does your depressed attitude prevent you from making an effort to accomplish something?

Gorse

____ Do you feel hopeless, as if there is no reason to try to improve things?

____ Do you lack faith that things could get better in your life and therefore make no effort to improve your circumstances?

____ Do you believe that nothing can be done to relieve your pain and suffering?

Heather

____ Do you find that others may avoid you because you seem to talk too much?

____ Do you dislike being alone, always seeking the companionship of others, to have someone to talk to?

____ Do your conversations usually wind up focusing on your interests or problems?

Holly

____ Are you suspicious of others, feeling that people have "ulterior motives"?

____ Do you feel great anger toward other people?

____ Are you full of jealousy, mistrust or hate?

Honeysuckle

____ Do you find yourself living in the past, nostalgic and homesick for the "way it was"?

____ Are you unable to change present circumstances because you are always looking back and never forward?

____ Do you often contemplate past regrets?

Hornbeam

____ Do you often feel too tired to face the day ahead?

____ Do you feel overworked or bored with your life?

____ Do you tend to procrastinate and put off some tasks while easily accomplishing those that are more enjoyable?

Impatiens

____ Do you feel a sense of urgency in everything you do, always rushing to get through things?

____ Are you impatient and irritable with others who seem to do things too slowly for you?

____ Do you prefer to work alone?

Larch

____ Do you lack self-confidence?

____ Do you feel inferior and often become discouraged?

____ Are you so sure that you will fail that you do not even attempt things?

Mimulus

____ Do you have fears of identifiable things, i.e. illness, death, pain heights, darkness, the dentist, etc?

____ Are you shy, overly sensitive and often afraid?

____ Do you often worry about everyday situations, i.e. traffic, bills, etc?

Mustard

____ Do you feel depressed without knowing why?

____ Do you feel your moods swinging back and forth?

____ Do you feel deep gloom, which seems to quickly appear for no apparent reason and then lifts just as suddenly?

Oak

____ Are you exhausted but feel the need to struggle on against all odds?

____ Do you have a strong sense of duty and dependability, carrying on no matter what obstacles stand in your way?

____ Do you neglect your own needs in order to complete a task?

Olive

____ Do you feel utterly and completely exhausted, both physically and mentally?

____ Are you totally drained of all energy with no reserves left, finding it difficult to carry on?

____ Have you just been through a long period of illness, stress, or strain with no relief?

Pine

____ Do you set overly high standards for yourself, never satisfied with your achievements?

____ Are you full of guilt and self-reproach?

____ Do you blame yourself for everything that goes wrong, sometimes even the mistakes of others?

Red Chestnut

____ Are you often concerned and worried about your loved ones?

____ Are you distressed and disturbed by other people's problems?

____ Do you worry that harm may come to those you care for?

Rock Rose

____ Are you susceptible to feeling of terror and panic?

____ Do you become helpless and frozen in the face of your fear?

____ Do you suffer from nightmares?

Rock Water

____ Do you set high personal standards and take pride in setting a good example for others?

____ Are you overly concerned with diet, exercise, work and spiritual disciplines?

____ Are you extremely disciplined in your approach to life, always striving for perfection?

Scleranthus

____ Do you find it difficult to decide when faced with a choice of two possibilities?

____ Do you lack concentration, always fidgety and nervous?

____ Do your moods change from one extreme to another: joy to sadness, optimism to pessimism, laughing to crying?

Star of Bethlehem

____ Have you suffered a recent shock in your life such as an accident, loss of a loved one, terrible news, illness?

____ Are you numbed or withdrawn because of traumatic events in your life?

____ Have you suffered a loss or grief from which you have never recovered?

Sweet Chestnut

____ Do you suffer from extreme mental or emotional anguish?

____ Do you feel that you have reached the limits of what you could possibly endure?

____ Do you feel as though there is no light at the end of the tunnel?

Vervain

____ Do you have so much energy and drive, that you're sometimes tense and can't fall asleep?

____ Do you have strong opinions and try to convince others of them?

____ Are you sensitive to injustice and dedicated to causes almost to the point that others think is extreme?

Vine

____ Do you tend to take charge of meetings, projects, situations, etc?

____ Do you consider yourself a natural leader?

____ Are you strong-willed and ambitious but may appear aggressive and domineering to others?

Walnut

____ Are you experiencing any change in you life - a move, new job, loss of someone loved, new relationship, divorce, puberty, menopause, giving up an addiction?

____ Do people or situations sometimes drain your energy?

____ Do you need to make a break from strong forces or attachments in you life that may be holding you back?

Water Violet

____ Do you appear to others to be aloof and overly proud?

____ Do you have a tendency to be withdrawn and prefer to be alone when faced with too many external distractions?

____ Do you bear your grief and sorrow without talking to others?

White Chestnut

____ Do you find your head full of persistent, unwanted thoughts that prevent concentration?

____ Do you relive unhappy events or arguments over and over again?

____ Are you unable to sleep at times because your mind seems to be cluttered with mental arguments that go round and round?

Wild Oat

____ Do you find yourself in a complete state of uncertainty over major life decisions?

____ Do you feel ready for a change of direction, but are unsure of which way to go?

____ Do you have ambition but feel that life is passing you by?

Wild Rose

____ Are you apathetic and resigned to whatever may happen in your life?

____ Do you have the attitude, "it doesn't matter anyhow"?

____ Do you lack the motivation to improve the quality of your life?

Willow

____ Do you feel resentful and bitter?

____ Do you have difficulty forgiving and forgetting?

____ Do you feel life is unfair and find yourself taking less interest in the things you used to enjoy?

♦

In the time since Dr. Bach's method spread worldwide, others have followed his methodology for discovering the applications of indigenous flowers in their respective regions. American, Australian, and Hawaiian flower essences have been researched, and they reiterate Dr. Bach's discovery that the flowering parts of plants have great impact on mental and emotional states.

Bibliography

- Bach, E.; *Health Thyself: An Explanation of the Real cause and Cure of Disease*, Random House UK, 1996
- Bach, E., Wheeler, E.; *The Bach Flower Remedies*, McGraw-Hill, 1998
- Craydon, D., Bellows, W.; *Floral Acupuncture: Applying the Flower Essences of Dr. Bach to Acupuncture Points*, Crossing Press, 2005
- Gotz, B.; *Advanced Bach Flower Therapy: A Scientific Approach to Diagnosis and Treatment*, Healing Arts Press; 1999

11

BIOLOGICAL MEDICINES

For our discussion here, biological medicines are defined as those made from organic substances or elements contained in them that have a therapeutic effect on a target organ or tissue in the body of the patient. Medically speaking, this is technically referred to as *organotherapy*.

In folk medicine, in organized botanical medicine, and also in homeopathic medicine, use has been made of what have traditionally been called "glandulars"—extracts of organic tissue taken orally in order to focus the attention of the healing power of the body on a particular organ. It has been observed over a long period of time that liver extract benefits the liver, for example. Some of these practices were adopted into orthodox medicine: Everyone is familiar with the use of thyroid gland extract to treat hypothyroidism. This is a prime example of a biological medicine. For decades, porcine or bovine thyroid tablets were the drug of choice until a synthetic thyroid came on the market.

However, where this is done in mainstream medicine, it is for the purpose of *substitutive* therapy. It is putting in from the outside what is deficient in the body. Without addressing why the substance may have been in short supply in the first place, a true cure never comes. Substitutive therapy is very popular and has full support of the pharmaceutical industry because one is a customer for life. As the organ becomes weaker due to the outside support, dependence on that support grows.

The tradition of eating organ meats according to one's weakness or illness is an old one; folk medicine in all cultures contains this concept. Many times a diet traditionally recommended for a particular malady has been found by modern science to contain the necessary elements needed to overcome the illness—even though the recipe originated long before any laboratory analysis of the nutritional contents of the foods.

The use of "glandulars" in healthcare—both folk medicine and organized medicine—is therefore well established. While they are often made from glands, they are not exclusively ductless or endocrine glands, and they are not given for their active hormone content. These tissue products are generally whole organs or tissues that are made by several different means. The oldest (and cheapest) method is dessication, which dries the ground-up tissue at high temperatures. This is not the optimal process; besides destroying all enzymes present in the tissue, it may destroy other biologically active substances. Salt-drying is also used, but this increases the sodium content considerably and makes them less universally well tolerated.

A more expensive but better process is freeze-drying. The low temperature preserves the enzymes and other factors present, and the fat is usually removed as well. This produces a more expensive product, but the end result is more near to consuming the fresh glandular. Technically, this product is said to be *lyophilized.*

The products are then ground up to make tablets or capsules. They lend support to the weak organ in a way that is different from substitutive therapy. Rather than providing the active hormone or other product of the gland or organ, thus taking over for it, "glandulars" stimulate the organ to modify its own functions, producing or regulating production.

Besides the aforemenioned processes, "glandulars" are also made by two other extraction processes, resulting in specific types of products. One is the *protomorphogenic* extract. This was developed in the 1940s by Dr. Royal Lee, a method of arriving at tissue extracts that contained what he called "cell determinants". These are structures that are the mineral framework onto which the chromosome of the cell is constructed. Along with associated nucleoproteins, the cell determinant forms organized groups which form the genes, which in turn form the chromosomes. They also form cell-mediated growth factors, and organize the structure of the cytoplasm of the cell. To put it simply, the cell determinants play a role in cell growth and regulation, and protomorphogens therefore represent a more concentrated and specific type of "glandular" product—nucleoproteins that can support repair of the target tissue or organ. The term Protomorphogen™ is now trademarked, as is the patented process of making it, by Standard Process, Inc.

The other process produces what is called a *cytosol* extract. This is a liquid extraction of the fluid within the cell and surrounding the nucleus. This fluid (the cytoplasm) is released when the cell walls of the glands or tissues are broken by the grinding process. Cytosol extracts supply the RNA and DNA information from the cell that causes the body to attract the necessary nutrients to repair damaged tissue in the target organ. They are not restricted to nucleoproteins found in the nucleus and therefore may contain other factors that may or may not be actively involved with the effects of the therapy. The principal company providing these products is Doctors' Research, Inc.

Why Tissue Extracts?

The practice of using tissue extracts to impact disease states is now appreciated in the age of autoimmune disease. Actually, many disease processes that have been with us for hundreds of years are now seen to be variations on the autoimmune process. The autoimmune reaction is a natural response that strives to keep stray nucleoproteins out of the bloodstream. That it is normally able to achieve this explains why we do not grow bone on the surface of our skin, or marrow in our eyes, etc. When the process is perverted and the person becomes in a sense allergic to his own tissue, selected tissue extracts will act as an oral antigen. The process is referred to as "oral tolerization", and is

125

considered to be one of the most exciting lines of research. A number of studies show that low-dose antigens—and remember the term *low dose*—are able to act against adenoviruses[38]. Oral tolerization is being studied as a possible answer to a number of human autoimmune diseases. In fact, even rejection of transplanted organs has been shown to be stemmed with the use of tissue extracts[39].

As far back as 1911, oral tolerization was shown to be effective. It was, interestingly enough, first demonstrated by researcher/writer H.G. Wells in England, who showed that feeding guinea pigs egg protein prevented severe and potentially lethal allergic reactions to injections of the same proteins. This should strike the aware reader as having implications involving homeopathic medicines, which we will mention shortly.

So at least part of the reason that tissue extracts have historically been effective in treating organic disease is because they act as a distraction for attacks on the organ by an autoimmune reaction, and they generate antibodies that go to work for the beleaguered organ. The simplistic philosophy of eating the body part that is ailing one may smack of magic on the surface, but modern science once again uncovers a hidden mechanism as to why such practices survive.

As to the matter of the low dose: It has been shown that it does not take a large material dose to exact an antigenic effect. In fact, a study of this method used in rheumatoid arthritis states "In animals, this phenomenon, termed oral tolerization, is strikingly dose-dependent in that only a narrow low-dose range produces protection."[40] While this means that small doses of

---

[38] Ilan, Y., et al.; Oral Tolerization To Adenoviral Antigens Permits Long-Term Gene Expression Using Recombinant Adenoviral Vectors *J. Clin. Invest.* 99(5): 1098-1106 (1997)

[39] Nagler, A., et al.; Oral Tolerization Ameliorates Liver Disorders Associated With Chronic Graft Versus Host Disease In Mice, *Hepatology* 2000; 31(3): 641-8

[40] Trentham, D.; Oral Tolerization As A Treatment Of Rheumatoid Arthritis, *Rheumatic Disease Clinics of North America*, 1998; 24, (3): 525-536

tissue extracts are safe and effective (particularly cytosol or pro-tomorphogenetic extracts that contain the concentrated genetic information of the type of tissue), it also suggests that there may be benefit in giving the same materials in microdoses, as is typical in homeopathic medicine. Later in this chapter, we will examine the major choices for use in this manner.

Organotherapy and Biological Medicines
in Psychoemotional Disorders

A number of organ tissue extracts can be used in the treatment of a wide variety of mental and emotional problems. Depending on the individual's presentation and the results of lab work, pituitary, adrenal and/or thyroid extract may be needed. The pituitary/adrenal/thyroid axis is central to balancing many endocrinological conditions that give rise to psychoemotional symptoms.

Some products are available containing brain tissue extracts (medulla, cerebrum, pituitary, hypothalamus, thalamus, pineal, etc.) that are specific brain cell activators. RNA and DNA are included in some, which act as "memorigenics". Co-factors such as amino acids and nutritive substances (covered next) enhance the actions of these brain "glandulars". There are some products on the market that couple these with botanicals for further augmentation of the therapeutic effect. In particular, the authors have been pleased with the efficacy of Serious Brain Enhancer™, Restful Mind Support™, and Simply Brain™, manufactured by Doctors' Research, Inc.[41] A glance at their ingredient lists shows that these formulations are of value in a wide variety of disorders.

Besides actual neurological and glandular tissue, the "resident" chemicals of those tissues, particularly neurotransmitters, are of use clinically as well. More than fifty neurotransmitters have been discovered, and many are involved with the experiencing of various emotions. The most widely studied (and manipu-

---

[41] 1248 E. Grand Avenue, Suite A, Arroyo Grande, CA 93420

lated) at present is serotonin. It regulates sleep-wake cycles, appetite, and emotional states. Moderate levels of serotonin inhibit anger and depression. A lowered level of this chemical in the hypothalamus and amygdala often results in these very states because of the loss of inhibitory function. Animal studies[42] have confirmed this, as aggressive animals have been found to have lower serotonin levels. Human subjects with depression generally present with low levels as well. Studies done using dietetic manipulation to reduce serotonin[43] showed that patients who were formerly depressed could be made to relapse by simply inhibiting the production of the chemical in the brain. Serotonin imbalance is one of the most frequent mechanisms for (note that we do not say "a cause of") stress reactions and depression.

Pedro L. Delgado, MD, Chairman of the Department of Psychiatry at University Hospitals of Cleveland and Case Western Reserve University, says "The unanswered question is: Why do some people become depressed when serotonin is low while others with low serotonin levels remain depression-free?"

While not a neurotransmitter, the essential amino acid Tryptophan does not occur naturally in the human body but is instead found in certain foods. When an individual consumes those foods, his or her body uses the tryptophan to produce serotonin.

Dopamine is another neurotransmitter that greatly affects mood. Motivation and self-drive will significantly lessen when dopamine levels are insufficient. Lowered efficacy and involvement with life are also experienced with imbalances of norepinephrine, and there will characteristically be low levels of physical energy. The person will not feel mentally motivated and will also not have the physical stamina to accomplish much even if she forces herself into action.

---

[42] Bernhardt, 1997

[43] Delgado, et al.; Serotonin Function and the Mechanism of Antidepressant Action: Reversal of Antidepressant Induce Remission by Rapid Depletion of Plasm Tryptophan, *Acrh Gen Psychiatry* 1990, 47:411-418

Acetylcholine is a neurotransmitter that acts both in the peripheral nervous system and the central nervous system. While it activates muscles in the peripheral nervous system, it is a neuromodulator in the central nervous system. This function is what concerns us here; acetylcholine plays a vital role in sensory perception, memory, and sustaining concentration.

Gamma-aminobutyric acid (GABA) is the chief inhibiting neurotransmitter in the central nervous system. It induces relaxation, reduces anxiety, and despite its calming action actually increases alertness. It is called the brain's "peacemaker chemical".

An imbalance of multiple neurotransmitters can result in depression, anxiety and panic attacks, sleep disorders, eating disorders, substance addiction, migraines and musculoskeletal pain.

Faulty metabolism, poor digestion and nutrient absorption, and simple nutritional deficiencies from faulty diet all play a role in creating varying neurotransmitter levels in the brain. Blood sugar fluctuations and insulin resistance have a powerful influence on neurotransmitter activity in the brain.

In the orthodox psychiatric approach, simply increasing a particular neurotransmitter's abundance is considered to have solved the problem. However, antidepressant drugs often seem to work for a while and then become ineffective, for several reasons. They have not been found to be effective overall, and the most often-prescribed category, selective serotonin reuptake inhibitor (SSRI) drugs, have no more activity than placebo in controlled trials[44]. After the expected relief is experienced, the placebo effect begins to wear off. For that lucky minority who initially benefits chemically from the drug, the "balance" created by increasing serotonin levels disappears, because no individualized balancing of the rest of the picture was created.

---

[44] *PLOS Med* 08; 5:e45

Now if we are to balance the various neurotransmitters in the way that a particular individual needs, we are closer to creating a real solution, and that is what some clinicians do. By ordering sophisticated tests (urine, saliva, and some blood tests) to arrive at the exact mix of brain chemicals present in the patient, a more comprehensive solution can be found. This more time-intensive, expensive method is a step up; however, once again simply mixing and matching brain chemicals is at best a "stop-gap measure". It may be good for stabilizing a case and buying time, but it does nothing to address the factors that caused the neurotransmitters to be imbalanced in the first place. Moreover, it ignores the capacity of the human body-mind system to put things right, if stimulated in the correct way. After all, those same neurotransmitters are produced there in the first place, are they not?

5-HTP

5-Hydroxytryptophan (5-HTP) is an amino acid that serves as a precursor to serotonin. It is also involved in tryptophan metabolism. It increases the synthesis and release of serotonin, which of course makes it theoretically valuable in conditions that are supposed to be caused by a lack of serotonin. Regardless of the accuracy of some of the theories about serotonin, 5-HTP has been effective in relieving depression, improving sleep, and regulating appetite. That it does in fact increase serotonin levels is evidenced by the fact that those taking selective serotonin reuptake inhibitor (SSRI) drugs must use caution in taking 5-HTP, as taking the two together can cause serotonin syndrome.

SAMe

S-adenosylmethionine (SAMe) is a nutritional supplement that is a synthetic form of a naturally occurring compound in the body, a combination of adenosine triphosphate and methionine. Clinical studies have shown SAMe to be effective for depression, as well as arthritis and liver disease. It is a considered a prescription drug in some European countries. This may be due to its side effects[45] but also its ability to cause hypomania and mania in bipolar patients. People with bipolar affective disorder and those with anxiety disorders should avoid this substance. Also, patients with Parkinson's disease taking Levodopa should avoid SAMe, as it could cause the Levodopa to lose its efficacy over time.

l-Tyrosine

l-Tyrosine, or simply Tyrosine, is an amino acid sometimes taken for depression and appetite control. It is made in the body from the amino acid phenylalanine. Tyrosine is converted into dopamine and norepinephrine, two of the major neurotransmitters that concern us. Supplemental tyrosine (and also phenylalanine) increases mental alertness and help concentration. Phenylalanine and tyrosine are sometimes prescribed together, usually in combination with other mood elevating nutrients and botanicals. Tyrosine side effects can occur with high doses, which places a limit on its applications.

One will often find in the literature a statement that "most people should not supplement with Tyrosine". While monitoring should be done if the person taking Tyrosine has phenylketonuria (PKU), its use has been generally safe. Note: No one with PKU should take phenylalanine.

Some human research with people suffering from a variety of conditions used 100 mg per 2.2 pounds of body weight, equivalent to about 7 grams per day for an average-sized person. The

---

[45] Nausea, dry mouth, hypoglycemic episodes, increased thirst, and anxiety.

appropriate amount to use in people with PKU is not known, therefore, the monitoring of blood levels by a physician is recommended.

l-Carnitine and Acetylcarnitine

l-Carnitine is another naturally occurring amino acid found in most cells of the body, including the brain. It is a common constituent of animal protein (meat, poultry, fish and also dairy products). Plant proteins have very small amounts[46]. It is taken as a supplement for improving mental clarity and mildly elevating the mood. Its activated form, acetylcarnitine (also spelled as acetyl-l-carnitine or l-acetylcarnitine), has a more dramatic impact on mental/emotional states, while carnitine appears to have more impact on the physical state, providing increased energy and endurance[47].

DMAE

DMAE (dimethyl-amino-ethanol) has been known by the product name Deanol in Europe for more than thirty years. DMAE is presently marketed as a dietary supplement, but at one time Deanol was used as a drug for hyperactivity in children. Deanol was thought to affect tardive dyskinesia (a trembling disorder caused by long-term anti-psychotic medication) because it was believed to be a cholinergic precursor that enhanced acetylcholine synthesis in the brain[48]. DMAE may in fact possess the ability to increase levels of the neurotransmitter acetylcholine, but studies that indicate that DMAE is a precursor to acetylcholine are not conclusive. It is still used, often successfully, for

---

[46] Non-vegetarians typically consume about 100 to 300 mg of carnitine daily, and the body is able to synthesize this nutrient in vegetarian diets if dietary intake is inadequate.

[47] Zanardi, R., Smeraldi, E.; A Double Blind Randomized Controlled Trial of Acetyl-l-carnitine vs. Amisulpride in the treatment of Dysthymia, *Eur Neuropsychopharmacol*, 2006:16(4): 281-7

[48] The results of studies using DMAE in tardive dyskinesia have been mostly negative.

132

Attention Deficit Disorder, enhancing memory and mood, boosting cognitive function, and increasing athletic performance.

Claims that DMAE is helpful in Alzheimer's disease, and age-related cognitive deficits, have not been substantiated. Topically, DMAE is used for signs of aging skin, particularly loose or sagging skin.

Homeopathic Tissue Remedies

Homeopathic medicine (Chapter Thirteen) has made use of organic extracts for many years, and the part of the materia medica given over to them is referred to as the tissue *sarcodes*. The tissue or organ desired (pancreas, heart muscle, fascia, etc.) is ground up, mixed, diluted and succussed to create an attenuated solution that carries the energetic "signature" of the source. They are given in this diluted form to target a healing response in the same tissue in the recipient's body (pituitary gland, etc.), or modulate the functions of a particular chemical (dopamine, etc.). Neurotransmitters have also been used in the homeopathic sarcode category.

The use of sarcodes is prominent in European approaches to Homeopathy, particularly in France. It is allied with what is usually referred to as "drainage therapy", operating on the theory that stimulation of an organ by specific remedies (or microdoses of its own tissue) will provoke a detoxification of that organ[49].

As a general rule, homeopathic sarcodes are given in lower potencies[50]; the lowest (3X up to 8X or 4C) is given when the target tissue or chemical is deficient and needs to be stimulated, and slightly higher (12X or 7C) for regulating and supporting, and a bit higher (18X or 9C) when it needs to be inhibited or modulated.

---

[49] Of course, the same dessicated or dehydrated organ extracts can be ingested in addition, and provide the exact amino acid profile needed as "building blocks" for the repair of the organ.

[50] The matter of homeopathic potencies is explained in Chapter 13.

Following is a symptom checklist, to better arrive at the neuro-transmitter most involved in the patient's problems. The higher the score in a particular block, the more likely that neurotransmitter needs to be targeted as part of therapy.

Acetylcholine	Dopamine
• Struggle to find words • Slower mental responses • Decreased memory • Difficulty visualizing things • Difficulty calculating • Decreased creativity • Different self-image • Excessive urination	• Feel worthless • Feel hopeless • Self-destructive thoughts • Easily irritated or angered • Decreased libido • Chronic infections • Inability to finish tasks • Unrefreshing sleep • Easily distracted • Loss of feeling for loved ones
GABA	Serotonin
• Feeling of impending doom • Panic or anxiety for no known reason • Can't shut mind off • Overwhelmed feeling • Agonizing over decisions • Can't focus on one subject	• Not enjoying life • Loss of enjoyment in previously enjoyable activities • Depressed feeling • Harder to fall asleep • Lowered pain threshold • Anger for no reason • Overcast weather makes mood worse

Bibliography

- Harrower, H., *Practical Organotherapy,* W.B. Conkey Co., Hammond, 1921
- Komaroff, A.; New Understanding of How Oral Tolerization Works, *Science* 1994 Aug 26; 265:1237-1240
- Lee, R., Hanson, W.; *Protomorpholgy*: *The Principles of Cell Auto-regulation*, Lee Foundation for Nutritional research, Milwaukee, 1947
- Maury, E.; *Drainage in Homeopathy*. Random House 1996

12

BOTANICAL MEDICINES

Botanical medicines, long a mainstay of naturopathic doctors, are the most scientifically scrutinized of all modalities used in natural medicine because they contain substantial amounts of pharmacologically active chemicals. One might be able to justify suspicion or outright disdain for dilute flower essences or homeopathic medicines, laugh at the theories of body-based psychotherapy, and sneer at the "pseudo-science" of muscle testing. Herbs? That's another thing altogether.

One-fourth of all prescription drugs are still derived from **plant sources.** Since these are prescription items, pharmaceutical companies are not anxious to advertise the fact that their often incredibly expensive drugs contain the same active ingredients as items one can buy cheaply in a health food store. Over 120 drugs listed in the United States Pharmacopoeia/National Formulary are plant-derived.

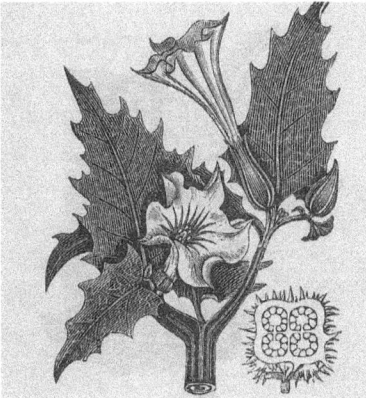

Around 80% of the people in developing countries still depend mostly on plant extracts for their medicine needs, according to the World Health Organization. But in the USA, regulations prohibit manufacturers from making claims for the products without lengthy and expensive testing. Therefore, the makers of most medicines that are totally derived from plant sources, whose manufacturers who are not conventional pharmaceutical companies, have a problem: They are not able to legally label their products with condition-specific indications. Because of this, many clinicians who investigate natural therapeutics gain a first impression that botanicals are at best a very poor cousin to synthesized chemical drugs. This is particularly untrue in prescribing for psychoemotional conditions.

Our recommendation is that botanical medicines be used as an initial measure, to stabilize the patient and provide a safe transition from synthetic drugs if they are being taken. As the case is taken in more detail, a more individualized plan can be had. Then, in combination with whichever other therapies are seen to be helpful, a prescription of a homeopathic medicine can be made that is more exact for the person—and more deeply corrective. This is not to say that botanical medicines are less effective; they are simply more adapted for acute and subacute stages, and they require less individualization in order to have an impact.

As will be detailed in the chapter on homeopathic medicines, the highly potentized homeopathic remedy has the potential to act on deep levels of disharmony in a way that material doses of plant-based extracts do not. On the other hand, the indicated homeopathic medicine is sometimes not reliable in an acute situation. As a simple explanation, it could be said that problems on the chemical level can be solved with chemicals, while problems on a more subtle level must be treated with subtle measures. Since we have accepted that mental and emotional illness cannot be simply put down to a misplaced molecule or two of a brain chemical, we must accept that a truly corrective therapy must reach to the core of the person's psyche where the problem has lodged[51]. That homeopathic medicines are able to do this is undeniable to anyone with experience in prescribing them. But they are so dependent on exactitude that they do not lend themselves to a high-volume practice. They can be largely

---

[51] Or solve situational issues in the person's life, if *that* is where the problem is lodged. For some people, only counseling and guidance will produce any resolution, no matter what oral agents they may or may not take.

ineffective if applied in a this-for-that fashion, which is exactly the way a beginning prescriber will try to use them.

Because of this, we advise the reader to match up the best botanical medicine (or medicines) for the case at hand, in order to "buy time" while gathering the necessary data that will result in a more precisely individualized prescription later. By perhaps applying other modalities as well (body-based psychotherapy, olfactory therapy, etc.), one can assess the value of these approaches and weed out those that are not having an effect, and ultimately coordinate a combined therapy program that will be more effective than any of the single therapies used.

The botanical medicines listed here are in the form of tinctures and fluid extracts. These liquid forms have superior efficacy and are more flexible in that they can be more easily combined to compound a medicine for a particular patient requiring a combination. In some cases where a liquid form is not typically available or advantageous, the solid form is given.

While it may seem contrary to what we have already expressed in this chapter and elsewhere about the importance of not being distracted by the pathology, the initial medicine list is categorized by condition (anxiety, depression of pregnancy, etc.), in order that one may more quickly find the possibly useful medicines to stabilize the patient. By then referring to the detailed description of a particular herb's indications, one can verify if the medicine applies to the patient or not. Whether the indicated botanical medicine is effective in stabilizing the person should be evident within a short time.

From that point, the more individualized medicine to unravel the patient's long-term problems can be sought by consulting the homeopathic repertory in Chapter Fifteen.

Index

Condition	Page	Condition	Page
Addiction withdrawal: alcohol	141	Depressive anxiety	174
Addiction withdrawal: drugs (See also **nightmares**)	143	Depression, endogenous	176
Addiction withdrawal: tobacco	144	Depression, menopausal	178
Anxiety	147	Depression, situational	180
Anxiety, menopausal	151	Depression, obsessive	182
Anxiety of pregnancy	153	Grief, despondency	183
Agitation due to over-work or exertion	154	Hyperactivity/ restlessness	186
Apathy, despondency	155	Hysteria	187
Asthenia, exhaustion	157	Labile emotions	190
Attention deficit disorder (ADD)	159	Nightmares and night terrors	190
Bipolar affective disorder: manic phase	165	Obsessive thinking	192
Bipolar affective disorder: depressive phase	168	Premenstrual syndrome (PMS)	193
Confusion	170	Post-traumatic stress disorder (PTSD)	196
Delusions	173	Seasonal affective disorder (SAD)	198

Key:
$\triangle$ P = use with caution during pregnancy
$\otimes$ P = contraindicated during pregnancy (Class 2b)
$\otimes$ N = contraindicated during nursing (Class 2c)
$\triangle$ = dosage must be strictly observed to avoid toxic effects (Class 3)
(Above in parenthesis are AHPA Botanical Safety Ratings)

Condition	Medicine	Indications
Addiction Withdrawal: Alcohol	*Apocynum can-nabinum* ⊗ P, △ Tincture: 1-5 gtt t.i.d.	Cathartic, car-diotonic, diuretic, and emetic. Calms stomach, strengthens heart during delirium tremens.
	Avena sativa Tincture: 10-50 gtt, t.i.d.	Antidepressive, sedative, stimu-lant, and tonic. Aids in overcom-ing alcohol habit.
	Capsicum anuum ⊗ P Tincture: 2-10 gtt.	Stimulant and anodyne. Contra-indicated orally in any gastrointesti-nal inflammation or ulceration.
	Chionanthus virginica △ Tincture: 5-10 gtt t.i.d.	Alterative, cho-lagogue, diuretic, hepatotonic, laxa-tive, tonic. Must use caution in biliary obstruc-tion.
	Datura stramo-nium △ Tincture: 1-5 gtt t.i.d.	Anodyne, anti-cholinergic, antispasmodic, sedative; Use in delirium tremens with rage, in-clined to violence (self or others).

Condition	Medicine	Indications
Addiction Withdrawal: Alcohol, cont'd.	*Gelsemium sempervirens* △, ⊗ P Tincture: 2-10 gtt, t.i.d.	Overdose can cause renal and respiratory failure. Frequent small doses can cause vomiting.
	Humulus lupulus Tincture: 30-50 gtt t.i.d., or infusion (1Tbsp dry herb to 1 cup hot H_2O), 2-4 oz t.i.d.	Delirium tremens, excitement, aids digestion.
	Hydrastis Canadensis ⊗P Tincture: 15-30 gtt t.i.d	To reduce craving.
	Lobelia inflata ⊗P, △ Tincture: 10-20 gtt q.i.d.	Antispasmodic, depressant, relaxant; Contraindicated in bradycardia.
	Panax quinquefolium △P Tincture: (cultivated) 10-50 gtt, t.i.d. (wild) 5-10 gtt, t.i.d. Infusion (1Tbsp dry herb to 1 cup hot H_2O): 2-4 oz t.i.d. Powdered herb: 10-60 gr.	Adaptogen, alterative, antispasmodic, cardiotonic, estrogenic, psychotonic, stimulant, stomachic, and tonic (esp. liver). Some authorities advise caution in pregnancy due to mild estrogenic effects.

Condition	Medicine	Indications
Addiction Withdrawal: Alcohol, cont'd.	*Passiflora incarnata* ⊗ P Tincture: 30-50 gtt up to q.i.d. Fl. Ext.: 30-60 gtt up to q.i.d. Powdered herb: 5-90 gr. up to q.i.d.	Sedative; Contraindicated in hypotension, bradycardia. Contraindicated in pregnancy due to presence of uterine stimulant (gynocardin).
	Scutellaria lateriflora ⊗ P Tincture: 20-60 gtt t.i.d. Infusion (1Tbsp dry herb to 1 cup hot H_2O): 2-6 oz t.i.d.	In delirium tremens, restlessness, insomnia, nightmares.
Addiction Withdrawal: Drugs (See also **nightmares**)	*Avena sativa* Tincture: 10-50 gtt, t.i.d.	Antidepressive, sedative, stimulant, tonic. Aids in overcoming drug habit.
	Datura stramonium △ Tincture: 1-5 gtt t.i.d.	Anodyne, anticholinergic, antispasmodic, sedative; Use in delirium tremens with rage, inclined to violence (self or others).
	Gelsemium sempervirens △, ⊗ P Tincture: 2-10 gtt, t.i.d.	Overdose can cause renal and respiratory failure. Frequent small doses can cause vomiting.

Condition	Medicine	Indications
Addiction Withdrawal: Drugs (See also **night-mares**), cont'd.	*Matriarca chamomilla* △ P Infusion (1Tbsp dry flowers to 1 cup hot H_2O), 2-6 oz PRN.	Caution in pregnancy if using parts of the plant other than the flowers, due to emmenagogue effects.
	Panax quinque-folium △ P Tincture: (culti-vated) 10-50 gtt, t.i.d. (wild) 5-10 gtt, t.i.d. Infusion (1Tbsp dry herb to 1 cup hot H_2O): 2-4 oz t.i.d. Powdered herb: 10-60 gr.	Adaptogen, al-terative, antispasmodic, cardiotonic, es-trogenic, psycho-tonic, stimulant, stomachic, and tonic (esp. liver) Some authorities advise caution in pregnancy due to mild estrogenic effects.
	Scutellaria lateriflora ⊗ P Tincture: 20-60 gtt t.i.d. Infusion (1Tbsp dry herb to 1 cup hot H_2O): 2-6 oz t.i.d.	In delirium tre-mens, restlessness, in-somnia, nightmares.
Addiction Withdrawal: Tobacco	*Acorus calamus* ⊗ P Tincture: 15-40 gtt, q.i.d.; Decoction: 1 tsp. herb to 1 cup H_2O Strain and drink a half cup to 1 cup, a.c.	Sedative, used for withdrawal symptoms. Con-traindicated in patients with pro-found emotional disorders or those taking MAO inhibitors.

Condition	Medicine	Indications
Addiction Withdrawal: Tobacco, cont'd.	*Avena sativa* Tincture: 10-50 gtt, t.i.d.	Antidepressive, Sedative, stimulant, tonic. Aids in overcoming habit.
	Glycyrrhiza glabra △P, △N (use de-glycyrrhizinated root) 5-60 gtt (or solid extract, ½ teaspoon), b.i.d.	Demulcent used for mucosal irritations. Glycyrrhiza is contraindicated in hypertension, and renal or cardiac insufficiency. De-glycyrrhizinated (DGL) licorice can be used relatively safely. Make sure patient is not potassium-deficient.
	Lobelia inflata ⊗P, △ Tincture: 10-20 gtt q.i.d.	Antispasmodic, bronchodilator, used for withdrawal and expectorant.
	Nepeta cataria Tincture: 10-60 gtt, t.i.d. Powdered herb: 5-45 gr. b.i.d.	For nervous irritability, agitation, weeping; patient drawing knees up is an indicating sign.

Condition	Medicine	Indications
Addiction Withdrawal: Tobacco, cont'd.	*Passiflora in-carnata* ⊗P Tincture: 30-50 gtt up to q.i.d. Fl. Ext.: 30-60 gtt up to q.i.d. Powdered herb: 5-90 gr. up to q.i.d.	Sedative; Contra-indicated in hypotension, bradycardia. Contraindicated in pregnancy due to presence of uterine stimulant (gynocardin).
	Scutellaria lateriflora ⊗P Tincture: 20-60 gtt t.i.d. Infusion (1Tbsp dry herb to 1 cup hot H_2O): 2-6 oz t.i.d.	In restlessness, insomnia, night-mares.
	Cola nitida ⊗P, △ Tincture: 5-25 gtt t.i.d., or infu-sion (1-2 tsp dry bark to 1 cup hot H_2O), 1-3 cups of the cold tea, q.d.	Tonic for depres-sive states, melancholic tem-perament. The patient is quiet and not conver-sant about troubles.
	Tussilago far-fara △P, △ Tincture: non-PA (pyrroliz-idine alkaloids) form,10-30 gtt t.i.d. Infusion (1-2 tsp. dry herb to 1 cup hot H_2O): 2-3 cups q.d.	Avoid prolonged use. Hepato-toxic.

Condition	Medicine	Indications
Anxiety	*Albizzia julibrissin* Tincture or Fl. Ext: 15-60 gtt, q.i.d.	A powerful mood elevator; useful for anxiety, fears, irritability, insomnia, nightmares, PTSD, moodiness, and disappointment. Do not use in manic patients.
	Anemone pulsatilla ⊗P, △N, △ Tincture: 1-10 gtt, t.i.d.	Sedative and antispasmodic, combines well with *Passiflora incarnata*.
	Avena sativa Tincture: 10-50 gtt, t.i.d.	Indicated in nervous exhaustion (neurasthenia) with anxiety, irritability, and labile emotions. Emotionally reactive to everything.
	Bacopa monnieri Tincture: 30-40 gtt, t.i.d. Powdered herb: 300 mg q.d.	Mental confusion with anxiety, and nervous exhaustion with agitation. Especially indicated for "cloudy thinking" and poor retention of new information. Contraindicated in hyperthyroidism.

Condition	Medicine	Indications
Anxiety, cont'd.	*Chamamaelum nobilis* ⊗P Tincture: 30-40 gtt, t.i.d.	Especially indicated in patients with weak digestion, flatulence, dysmenorrhea.
	Eschscholtzia californica ⊗P Tincture: 10-30 gtt, q.i.d.	Indicated in patients with anxiety, restless insomnia, nervous tension, and stress headaches.
	Humulus lupulus Tincture: 30-50 gtt t.i.d., or infusion (1 Tbsp dry herb to 1 cup hot H_2O), 2-4 oz t.i.d.	Mild anxiety, nervous irritability, and wakefulness, possibly with gastric symptoms.
	Leonurus cardiaca △P Tincture: 10-30 gtt, t.i.d	Mild sedative and anxiolytic. Combine with *Verbena officinalis* for PMS, menstrual or menopausal anxiety. Add *Anemone pulsatilla* for panic attacks.

Condition	Medicine	Indications
Anxiety, cont'd.	*Lilium tigrinum* △ Tincture: 1-5 gtt, t.i.d. Use in moderation.	For anxiety with depression. Weeping, muttering under the breath, and fear of solitude are indicators. Often used in menstrual, menopausal or post-partum depression. Overdose can cause saponin toxicity, leading to somnolence, vomiting and purging.
	Matriarca chamomilla △P Infusion (1Tbsp dry flowers to 1 cup hot H_2O), 2-6 oz PRN.	Caution in pregnancy if using parts of the plant other than the flowers, due to emmenagogue effects.
	Passiflora incarnata ⊗P Tincture: 30-50 gtt up to q.i.d. Fl. Ext.: 30-60 gtt up to q.i.d. Powdered herb: 5-90 gr. up to q.i.d.	Specific for nervous restlessness; sleeplessness with muscle twitching, or circular thinking.

Condition	Medicine	Indications
Anxiety, cont'd.	*Piper methysti-cum* ⊗P Tincture: 5-60 gtt t.i.d. Fl. Ext.: 5-60 gtt, t.i.d. Powdered herb: 14 gr (2 #OO capsules) t.i.d.	For anxiety with muscle tension, bruxism, restless leg syndrome, and pain.
	Primula veris Tincture: 15-20 gtt, t.i.d.	Anxiety states associated with restlessness and irritability; combines well with *Scutellaria lateri-flora.*
	Valeriana offici-nalis △ Tincture: 30-50 gtt, t.i.d. (tea-spoonful doses may be required for acute use) Fl. Ext.: 5-30 gtt Powdered herb: 2 g (2 #OO cap-sules), t.i.d.	Indicated for nervous, restless patient, with agi-tation, a pale face and clammy skin. Note: protracted use of dried herb can increase anxiety and/or become a nerv-ous depressant. May potentiate effects of CNS depressant drugs.

Condition	Medicine	Indications
Anxiety, menopausal	*Leonurus cardiaca* Δ P Tincture: 10-30 gtt, t.i.d	Mild sedative and anxiolytic. Combine with *Verbena officinalis* for PMS, menstrual or menopausal anxiety. Add *Anemone pulsatilla* for panic attacks.
	Lilium tigrinum Δ Tincture: 1-5 gtt, t.i.d. Use in moderation.	For anxiety with depression. Weeping, muttering under the breath, and fear of solitude are indicators. Often used in menstrual, menopausal or post-partum depression. Overdose can cause saponin toxicity, leading to somnolence, vomiting and purging.

Condition	Medicine	Indications
Anxiety, menopausal, cont'd.	*Selenicereus grandiflorus* △ Tincture: 5-30 gtt, t.i.d.	Especially indicated for anxiety associated with cardiac conditions. Indications: Irregular pulse, feeble or violent heart action, sensation of constriction in the chest (Note: Contraindicated in acute onset, with tachycardia or hypertension. Caution using in patients taking beta or calcium channel blockers).
	Scutellaria lateriflora ⊗ P Tincture: 20-60 gtt t.i.d. Infusion (1Tbsp dry herb to 1 cup hot H_2O): 2-6 oz t.i.d.	Nervousness, twitching, or spasms due to mental or physical exertion. Nervousness without apparent cause. Anger, irritability, angry outbursts.
	Tilia platyphyllos Tincture: 25-50 gtt, t.i.d.	Sedative but also tonic; combines well with *Humulus lupulus*. Frequent use not advised in unstable cardiac patients.

Condition	Medicine	Indications
Anxiety, menopausal, cont'd.	*Verbena offici-nalis* △ P Tincture: 15-20 gtt, t.i.d.	Menopausal anxi-ety, anxiety with nervous tics, tremors, or spasms. Com-bines well with *Leonurus* and *Pulsatilla*.
Anxiety of pregnancy	*Turnera diffusa* △ P Tincture: 10-30 gtt, t.i.d. Fl. Ext.: ½ to 1 dr. Powdered herb: 15-45 gr.	For anxiety neu-rosis, usually with predominantly sexual factor.
	Viburnum opulus Tincture: 30-60 gtt up to q.i.d. Fl. Ext.: 15-40 gtt, t.i.d. Infusion (1 oz herb to 1 pint H_2O): 1 cup t.i.d.	For anxiety asso-ciated with pregnancy. Com-bines well with *Avena sativa*. Note: Do not con-fuse with *Viburnum prunifo-lium*, which is similar in action but contraindi-cated in pregnancy.

Condition	Medicine	Indications
Agitation due to over-work or exertion	*Anemone pul-satilla* ⊗ P, △ N, △ Tincture: 1-10 gtt, t.i.d.	Especially with insomnia; com-bines well with *Passiflora incar-nata.*
	Eschscholtzia californica ⊗ P Tincture: 10-30 gtt, q.i.d.	Indicated in pa-tients with anxiety, restless insom-nia, nervous tension, and stress head-aches.
	Passiflora incarnata ⊗ P Tincture: 30-50 gtt up to q.i.d. Fl. Ext.: 30-60 gtt up to q.i.d. Powdered herb: 5-90 gr. up to q.i.d.	Irritability, insom-nia from worry, overwork.
	Piper methysti-cum ⊗ P Tincture: 5-60 gtt t.i.d. Fl. Ext.: 5-60 gtt, t.i.d. Powdered herb: 14 gr (2 #OO capsules) t.i.d.	For anxiety with muscle tension, bruxism, restless leg syndrome, and pain.

Condition	Medicine	Indications
Agitation due to over-work or exertion, cont'd.	*Scutellaria lateriflora* ⊗ P Tincture: 20-60 gtt t.i.d. Infusion (1Tbsp dry herb to 1 cup hot H_2O): 2-6 oz t.i.d.	Nervousness, twitching, or spasms due to mental or physi-cal exertion. Nervousness without apparent cause. Anger, irritability, angry outbursts.
Apathy, despondency	*Baptisia tincto-ria* ⊗ P Tincture: 10-20 gtt, t.i.d.	Although usually employed as an antimicrobial, *Baptisia* exerts an influence on this mental state. Large doses are toxic; not for long-term use.
	Cimicifuga ra-cemosa ⊗ P, △ N, △ Tincture: 30-90 gtt, t.i.d. Fl. Ext: 5-30 gtt.	Indicated for "gloom and doom" depres-sion, hormonal depression.
	Cola nitida ⊗ P, △ Tincture: 5-25 gtt t.i.d., or infu-sion (1-2 tsp dry bark to 1 cup hot H_2O), 1-3 cups of the cold tea, q.d.	Indicated in nerv-ous exhaustion with mental de-spondency, foreboding. The patient is quiet and not conver-sant about troubles.

Condition	Medicine	Indications
Apathy, despondency, cont'd.	*Lavandula angustifolia* △ P Tincture: 25-50 gtt, t.i.d. Infusion: 1 tsp lavender flowers to 1 cup hot H_2O), steep 10 minutes. 1 cup q.i.d. Note: Essential <u>oil</u> of lavender is toxic and contraindicated in pregnancy.	Indicated for restlessness, poor sleep, foggy, difficult thinking; nervous strain affecting the stomach and intestines.
	Oenothera biennis △ (Evening Primrose oil) Oil: 1-3 tsp q.d. OO capsules: 2-6 q.d.	Indicated in depression associated with chronic dyspepsia, nausea, and urinary frequency. Patient is typically apathetic, gloomy, and despondent.
	Valeriana officinalis △ Tincture: 30-50 gtt, t.i.d. (teaspoonful doses may be required for acute use) Fl. Ext.: 5-30 gtt Powdered herb: 2 g (2 #OO capsules), t.i.d.	Indicated for nervous, restless patient, with agitation, a pale face and clammy skin. Note: protracted use of dried herb can increase anxiety and/or become a nervous depressant. May potentiate effects of CNS depressant drugs.

Condition	Medicine	Indications
Asthenia, exhaustion	*Avena sativa* Tincture: 10-50 gtt, t.i.d.	Indicated in nervous exhaustion (neurasthenia) with anxiety, irritability, and labile emotions. Emotionally reactive to everything.
	Bacopa monnieri Tincture: 30-40 gtt, t.i.d. Powdered herb: 300 mg q.d.	Mental confusion with anxiety, and nervous exhaustion with agitation. Especially indicated for "cloudy thinking" and poor retention of new information. Contraindicated in hyperthyroidism.
	Cimicifuga racemosa ⊗P, △N, △ Tincture: 30-90 gtt, t.i.d. Fl. Ext: 5-30 gtt.	Indicated for "gloom and doom" depression, hormonal depression.
	Cola nitida ⊗P, △ Tincture: 5-25 gtt t.i.d., or infusion (1-2 tsp dry bark to 1 cup hot H_2O), 1-3 cups of the cold tea, q.d.	Indicated in nervous exhaustion with mental despondency, foreboding. The patient is quiet and not conversant about troubles.

Condition	Medicine	Indications
Asthenia, exhaustion, cont'd.	*Cordyceps sinensis* 500 mg q.d. or b.i.d.	Cardiotonic, hepatoprotective, immunostimulant, stomachic, and tonic effects.
	Eleutherococcus senticosus Tincture 20-60 gtt, t.i.d. Infusion (1Tbsp dry herb to 1 cup hot H_2O): 2-4 oz, t.i.d.	Adaptogen, adrenergic, anti-depressive, hypertensive, tonic, vasodilator.
	Glycyrrhiza glabra △P, △N, △ Tincture: (use de-glycyrrhizinated root) 5-60 gtt (or solid extract, ½ teaspoon), b.i.d.	Glycyrrhiza is contraindicated in hypertension, and renal or cardiac insufficiency. Deglycyrrhizinated (DGL) licorice can be used relatively safely. Make sure patient is not potassium-deficient.
	Medicago sativum △P Tincture: 5-30 gtt, q.i.d. Infusion of dried herb (1 Tbsp leaves to 1 cup hot H_2O): 2-6 oz t.i.d.	Estrogenic (mild), nutritive, tonic.

Condition	Medicine	Indications
Asthenia, exhaustion, cont'd.	*Panax quinque-folium* △ P Tincture: (culti-vated) 10-50 gtt, t.i.d. (wild) 5-10 gtt, t.i.d. Infusion (1Tbsp dry herb to 1 cup hot H_2O): 2-4 oz t.i.d. Powdered herb: 10-60 gr.	Adaptogen, al-terative, antispasmodic, cardiotonic, es-trogenic, psychotonic, stimulant, stom-achic, and tonic (esp. liver). Some authorities advise caution in preg-nancy due to mild estrogenic ef-fects.
	Withania som-nifera Tincture: 10-30 gtt, t.i.d. Decoction (1 Tbsp of dried herb to a cup of H_2O, boil 4 min-utes, steep 10 minutes): Strain and drink 1 cup t.i.d.	Adaptogen and tonic.
Attention Deficit Disorder	*Avena sativa* Tincture: 10-50 gtt, t.i.d.	Dose for one month and evalu-ate.

Condition	Medicine	Indications
Attention Deficit Disorder, cont'd.	*Atropa bella-donna* △ Best given in 3X (1:1000) attenuation (homeopathic dose); Tincture can be added to 4 oz. H₂O and ½ teaspoon doses given t.i.d.: Tincture (leaf) 2-5 gtt Tincture (root) 1 gt.	Indicated in nervous excitement, hysterical presentation with delirium, dilated pupils.
	Ferula asafetida ⊗P Tincture: 5-30 gtt t.i.d.	Depression with nervous irritation, possibly hysteria, headache, dizziness, and flatulence. Tight muscles; may be ticklish, or painful to the touch.
	Gelsemium sempervirens △, ⊗ P Tincture: 2-10 gtt, t.i.d. Fl. Ext.: 1-5 gtt, t.i.d.	Indications: hot head, flushed face, bright eyes, restlessness and agitation.
	Cimicifuga racemosa ⊗P, △N, △ Tincture: 30-90 gtt, t.i.d. Fl. Ext: 5-30 gtt.	Indicated for "gloom and doom" depression, hormonal depression.

Condition	Medicine	Indications
Attention Deficit Disorder, cont'd.	*Datura stramonium* △ Tincture: 1-5 gtt t.i.d.	Indicated in alternating fits of weeping and laughter.
	Eleutherococcus senticosus Tincture 20-60 gtt, t.i.d. Infusion (1Tbsp dry herb to 1 cup hot H_2O): 2-4 oz, t.i.d.	Adaptogen, adrenergic, antidepressive, hypertensive, tonic, vasodilator.
	Eschscholtzia californica ⊗P Tincture: 10-30 gtt, q.i.d.	Indicated in patients with anxiety, restless insomnia, nervous tension, and stress headaches.
	Oenothera biennis △ (Evening Primrose oil) Oil: 1-3 tsp q.d. OO capsules: 2-6 q.d.	Indicated in depression associated with chronic dyspepsia, nausea, and urinary frequency. Patient is typically apathetic, gloomy, and despondent.
	Humulus lupulus Tincture: 30-50 gtt t.i.d., or infusion (1Tbsp dry herb to 1 cup hot H_2O), 2-4 oz t.i.d.	Mild anxiety, nervous irritability, and wakefulness, possibly with gastric symptoms.

Condition	Medicine	Indications
Attention Deficit Disorder, cont'd.	*Hypericum perforatum* △ P, △ N, △ Fl.Ex.: 5 gtt, b.i.d. for young children under 5 y.o.; 10 gtt up to t.i.d. for school age. Infusion: 1 tsp cut herb in 200 ml of boiling water; steep for 10 minutes; Daily dosage:1 cup for young children, 2-3 cups for children over 3 years old.	Has mild sedative action, especially with nervous restlessness and sleep disorders.
	Lavandula angustifolia △ P Tincture: 25-50 gtt, t.i.d. Infusion: 1 tsp lavender flowers to 1 cup hot H_2O), steep 10 minutes. 1 cup q.i.d. Note: Essential oil of lavender is toxic and contraindicated in pregnancy.	Indicated for restlessness, poor sleep, foggy, difficult thinking; nervous strain affecting the stomach and intestines.

Condition	Medicine	Indications
Attention Deficit Disorder, cont'd.	*Matriarca chamomilla* △ P Infusion (1Tbsp dry flowers to 1 cup hot H_2O), 2-6 oz PRN.	Caution in pregnancy if using parts of the plant other than the flowers, due to emmenagogue effects.
	Melissa officinalis △ P Tincture: 30-50 gtt t.i.d. Infusion (1 Tbsp dry herb to a cup of H_2O): 1 cup q1-2h PRN for pain.	Antidepressant, antispasmodic, sedative. Insomnia due to nervous strain; gastrointestinal problems. Contraindicated in hypothyroid patients. Lower doses in pregnancy.
	Passiflora incarnata ⊗ P Tincture: 30-50 gtt up to q.i.d. Fl. Ext.: 30-60 gtt up to q.i.d. Powdered herb: 5-90 gr. up to q.i.d.	Specific for nervous restlessness; sleeplessness with muscle twitching, or circular thinking.

Condition	Medicine	Indications
Attention Deficit Disorder, cont'd.	*Rhodiola rosea* Fl. Ext.: 5-40 gtt a.c., t.i.d. Powdered root: 1 g a.c., t.i.d. standardized extract (3% rosavins and 0.8 to 1% salidroside) 200-600 mg a.c., q.d. Lower dosage is stimulating, higher dosage is sedating. 800 mg q.d. on empty stomach is the usual dose for hyperactivity.	
	Scutellaria lateriflora ⊗P Tincture: 20-60 gtt t.i.d. Infusion (1Tbsp dry herb to 1 cup hot H_2O): 2-6 oz t.i.d. Note: This herb should not be given to a child younger than six y.o.	Relaxant; calms the mind.
	Datura stramonium △ Tincture: 1-5 gtt t.i.d.	Indicated in alternating fits of weeping and laughter.

Condition	Medicine	Indications
Attention Deficit Disorder, cont'd.	*Eschscholtzia californica* ⊗ P Tincture: 10-30 gtt, q.i.d.	Indicated in patients with anxiety, restless insomnia, nervous tension, and stress headaches.
Bipolar affective disorder: manic phase	*Humulus lupulus* Tincture: 30-50 gtt t.i.d., or infusion (1Tbsp dry herb to 1 cup hot H_2O), 2-4 oz t.i.d.	Mild anxiety, nervous irritability, and wakefulness, possibly with gastric symptoms.
	Hypericum perforatum △ P, △ N, △ Fl.Ex.: 5 gtt, b.i.d. for young children under 5 y.o.; 10 gtt up to t.i.d. for school age. Infusion: 1 tsp cut herb in 200 ml of boiling water; steep for 10 minutes; Daily dosage:1 cup for young children, 2-3 cups for children over 3 years old.	Has mild sedative action, especially with nervous restlessness and sleep disorders.

Condition	Medicine	Indications
Bipolar affective disorder: manic phase, cont'd.	*Lavandula angustifolia* △ P Tincture: 25-50 gtt, t.i.d. Infusion: 1 tsp lavender flowers to 1 cup hot H$_2$O), steep 10 minutes. 1 cup q.i.d. Note: Essential oil of lavender is toxic and contraindicated in pregnancy.	Indicated for restlessness, poor sleep, foggy, difficult thinking; nervous strain affecting the stomach and intestines.
	Matriarca chamomilla △ P Infusion (1Tbsp dry flowers to 1 cup hot H$_2$O), 2-6 oz PRN.	Caution in pregnancy if using parts of the plant other than the flowers, due to emmenagogue effects.
	Melissa officinalis △ P Tincture: 30-50 gtt t.i.d. Infusion (1 Tbsp dry herb to a cup of H$_2$O): 1 cup q1-2h PRN for pain.	Antidepressant, antispasmodic, sedative. Insomnia due to nervous strain; gastrointestinal problems. Contraindicated in hypothyroid patients. Lower doses in pregnancy.

Condition	Medicine	Indications
Bipolar affective disorder: manic phase, cont'd.	*Passiflora incarnata* ⊗P Tincture: 30-50 gtt up to q.i.d. Fl. Ext.: 30-60 gtt up to q.i.d. Powdered herb: 5-90 gr. up to q.i.d.	Specific for nervous restlessness; sleeplessness with muscle twitching, or circular thinking.
	Piper methysticum ⊗P Tincture: 5-60 gtt t.i.d. Fl. Ext.: 5-60 gtt, t.i.d. Powdered herb: 14 gr (2 #OO capsules) t.i.d.	For anxiety with muscle tension, bruxism, restless leg syndrome, and pain.
	Rhodiola rosea Fl. Ext.: 5-40 gtt a.c., t.i.d. Powdered root: 1 g a.c., t.i.d. standardized extract (3% rosavins and 0.8 to 1% salidroside) 200-600 mg a.c., q.d. Lower dosage is stimulating, higher dosage is sedating. 800 mg q.d. on empty stomach is the usual dose for mania.	

Condition	Medicine	Indications
Bipolar affective disorder: manic phase, cont'd.	*Avena sativa* Tincture: 10-50 gtt, t.i.d.	Tonic effects with antidepressant activity.
	Centella asiatica △ P Tincture: 20-40 gtt, t.i.d.	Contraindicated in hyperthyroidism and myxedema. Caution in pregnancy due to mild emmenagogue and abortifacient effects.
Bipolar affective disorder: depressive phase	*Melissa officinalis* △ P Tincture: 30-50 gtt t.i.d. Infusion (1 Tbsp dry herb to a cup of H_2O): 1 cup q1-2h PRN for pain.	A mild mood elevator. Use in combination with *Hypericum*. Antidepressant, antispasmodic, and sedative. Insomnia due to nervous strain; gastrointestinal problems. Contraindicated in hypothyroid patients. Lower doses in pregnancy.
	Scutellaria lateriflora ⊗ P Tincture: 20-60 gtt t.i.d. Infusion (1Tbsp dry herb to 1 cup hot H_2O): 2-6 oz t.i.d.	Nervousness, twitching, or spasms due to mental or physical exertion. Nervousness without apparent cause. Anger, irritability, angry outbursts.

Condition	Medicine	Indications
Bipolar affective disorder: depressive phase, cont'd.	*Turnera diffusa* Δ P Tincture: 10-30 gtt, t.i.d. Fl. Ext.: ½ to 1 dr. Powdered herb: 15-45 gr.	For mild depression with a marked loss of libido; or anxiety, usually with predominantly sexual factor.
	Bacopa monnieri Tincture: 30-40 gtt, t.i.d. Powdered herb: 300 mg q.d.	Mental confusion with anxiety, and nervous exhaustion with agitation. Especially indicated for "cloudy thinking" and poor retention of new information. Contraindicated in hyperthyroidism.
	Centella asiatica Δ P Tincture: 20-40 gtt, t.i.d.	Improves cerebral circulation. Contraindicated in hyperthyroidism and myxedema. Caution in pregnancy due to mild emmenagogue and abortifacient effects.

Condition	Medicine	Indications
Confusion	*Ginkgo biloba* △ Tincture: 30-60 gtt, up to t.i.d. (or 250 mg of dried herb q.d.).	Improves cerebral function; indicated in confusion caused by vascular insufficiency due to old age, or head trauma. Caution: May decrease efficacy of anticonvulsant drugs, Omeprazole, and Tolbutamide. May increase antiplatelet activity of blood-thinning drugs. May potentiate effects of Haloperidol. May increase levels of Nifedipine, Digoxin, and beta-blockers. May cause hypoglycemia in patients taking oral hypoglycemic drugs (Glipizide, etc.).

Condition	Medicine	Indications
Confusion, cont'd.	*Melissa offici-nalis* △ **P** Tincture: 30-50 gtt t.i.d. Infusion (1 Tbsp dry herb to a cup of H_2O): 1 cup q1-2h PRN for pain.	A mild mood elevator. Use in combination with *Hypericum*. Anti-depressant, antispasmodic, and sedative. In-somnia due to nervous strain; gastrointestinal problems. Con-traindicated in hypothyroid pa-tients. Lower doses in preg-nancy.
	Passiflora in-carnata ⊗ **P** Tincture: 30-50 gtt up to q.i.d. Fl. Ext.: 30-60 gtt up to q.i.d. Powdered herb: 5-90 gr. up to q.i.d.	Specific for nerv-ous restlessness; sleeplessness with muscle twitching, or cir-cular thinking.
	Rhodiola rosea Fl. Ext.: 5-40 gtt a.c., t.i.d. Powdered root: 1 g a.c., t.i.d. Standardized extract (3% ro-savins and 0.8 to 1% slidro-side) 200-600 mg a.c., q.d. Lower dosage is stimulating, higher dosage is sedating.	

Condition	Medicine	Indications
Confusion, cont'd.	*Rosmarinus officinalis* △ P Tincture: 25-50 gtt, t.i.d. Infusion (1 oz herb to 1 pint H_2O): 2-4 oz t.i.d. Oil: 3-6 gtt in a binding vehicle, and taken with H_2O.	Indicated in dull, lethargic depression; patient doesn't want to; mentally "foggy".
	Physostigma venenosum △ Tincture: 5 gtt per day.	
	Scutellaria lateriflora ⊗ P Tincture: 20-60 gtt t.i.d. Infusion (1 Tbsp dry herb to 1 cup hot H_2O): 2-6 oz t.i.d. Note: This herb should not be given to a child younger than six y.o.	Relaxant; calms the mind. Indicated in nervousness without apparent cause. Anger, irritability, angry outbursts.

Condition	Medicine	Indications
Delusions	*Valeriana offici-nalis* Δ Tincture: 30-50 gtt, t.i.d. (tea-spoonful doses may be required for acute use) Fl. Ext.: 5-30 gtt Powdered herb: 2 g (2 #OO cap-sules), t.i.d.	Indicated for nervous, restless patient, with agi-tation, a pale face and clammy skin. Imagined worries. Note: protracted use of dried herb can increase anxiety and/or become a nerv-ous depressant. May potentiate effects of CNS depressant drugs.
	Ferula asafetida ⊗P Tincture: 5-30 gtt t.i.d.	Depression with nervous irritation, possibly hysteria, headache, dizzi-ness, and flatulence. Tight muscles; may be ticklish, or painful to the touch.
	Lilium tigrinum Δ Tincture: 1-5 gtt, t.i.d. Use in modera-tion.	For anxiety with depression. Weeping, mutter-ing under the breath, and fear of solitude are indicators. Often used in menstru-al, menopausal or post-partum de-pression. Over-dose can cause saponin toxicity, leading to somno-lence, vomiting and purging.

Condition	Medicine	Indications
Depressive anxiety	*Rhodiola rosea* Fl. Ext.: 5-10 gtt a.c., t.i.d. Powdered root: 500 mg, t.i.d. standardized extract (3% rosavins and 0.8 to 1% salidroside) 200 mg a.c., q.d.	Tonic.
	Selenicereus grandiflorus Δ Tincture: 5-30 gtt, t.i.d.	Especially indicated for anxiety associated with cardiac conditions. Indications: Irregular pulse, feeble or violent heart action, sensation of constriction in the chest (Note: Contraindicated in acute onset, with tachycardia or hypertension. Caution using in patients taking beta or calcium channel blockers).

Condition	Medicine	Indications
Depressive anxiety, cont'd.	*Scutellaria lateriflora* ⊗ P Tincture: 20-60 gtt t.i.d. Infusion (1Tbsp dry herb to 1 cup hot H$_2$O): 2-6 oz t.i.d.	Nervousness, twitching, or spasms due to mental or physical exertion. Nervousness without apparent cause. Anger, irritability, angry outbursts.
	Albizzia julibrissin Tincture or Fl. Ext: 15-60 gtt, q.i.d.	A powerful mood elevator; useful for anxiety, fears, irritability, insomnia, nightmares, PTSD, moodiness, and disappointment. Do not use in manic patients.
	Cimicifuga racemosa ⊗ P, △ N, △ Tincture: 30-90 gtt, t.i.d. Fl. Ext: 5-30 gtt.	Indicated for "gloom and doom" depression, hormonal depression.

Condition	Medicine	Indications
Depression, endogenous	*Hypericum per-foratum* △ P, △ N, △ Tincture: 5-30 gtt (or 200-400 mg dried herb), t.i.d. for a short time, then decrease dose to b.i.d. Antidepressant effects require substantial doses (300-400 mg b.i.d.) and take 2-3 weeks to manifest.	For mild to moderate unipolar or situational depression. "Sour stomach + sour attitude" is a common feature. Large doses can create photosensitivity. Use caution in patients undergoing heliotherapy or ultraviolet therapy.
	Ocimum sanctum Tincture: 30-50 gtt t.i.d. Infusion (1 Tbsp dry herb to 1 cup hot H_2O): 2-6 oz t.i.d.	Indicated when the patient is fixated on a specific topic or event. Usually fatigued and mentally foggy. Combines well with *Lavandula, Hypericum,* and *Rosmarinus.*
	Oenothera bi-ennis △ (Evening Primrose oil) Oil: 1-3 tsp q.d. OO capsules: 2-6 q.d.	Indicated in depression associated with chronic dyspepsia, nausea, and urinary frequency. Patient is typically apathetic, gloomy, and despondent.

Condition	Medicine	Indications
Depression, endogenous, cont'd.	*Rosmarinus officinalis* △P Tincture: 25-50 gtt, t.i.d. Infusion (1 oz herb to 1 pint H_2O): 2-4 oz t.i.d. Oil: 3-6 gtt in a binding vehicle, and taken with H_2O.	Indicated in dull, lethargic depression; patient doesn't want to; mentally "foggy".
	Turnera diffusa △P Tincture: 10-30 gtt, t.i.d. Fl. Ext.: ½ to 1 dr. Powdered herb: 15-45 gr.	For anxiety neurosis, usually with predominantly sexual factor.
	Valeriana officinalis △ Tincture: 30-50 gtt, t.i.d. (teaspoonful doses may be required for acute use) Fl. Ext.: 5-30 gtt Powdered herb: 2 g (2 #OO capsules), t.i.d.	Indicated for nervous, restless patient, with agitation, a pale face and clammy skin. Note: protracted use of dried herb can increase anxiety and/or become a nervous depressant. May potentiate effects of CNS depressant drugs.

Condition	Medicine	Indications
Depression, endogenous, cont'd.	*Actea alba* ⊗ P Tincture: 1-20 gtt, t.i.d. Usual dose is 20 gtt mixed in 4 oz H_2O, 1 tsp q3h.	Antispasmodic, anticonvulsive, antidepressant; specifically indicated in tenderness in the ovarian region.
	Cimicifuga racemosa ⊗ P, △ N, △ Tincture: 30-90 gtt, t.i.d. Fl. Ext: 5-30 gtt.	Indicated for " gloom and doom " depression, hormonal depression.
Depression, menopausal	*Leonurus cardiaca* △ P Tincture: 10-30 gtt, t.i.d	Mild sedative and anxiolytic. Combine with *Verbena officinalis* for PMS, menstrual or menopausal anxiety. Add *Anemone pulsatilla* for panic attacks.
	Lilium tigrinum △ Tincture: 1-5 gtt, t.i.d. Use in moderation.	For anxiety with depression. Weeping, muttering under the breath, and fear of solitude are indicators. Often used in menstrual, menopausal or post-partum depression. Overdose can cause saponin toxicity, leading to somnolence, vomiting and purging.

Condition	Medicine	Indications
Depression, menopausal, cont'd.	*Verbena offici- nalis* △ P Tincture: 15-20 gtt, t.i.d.	Menopausal anxi- ety, anxiety with nervous tics, tremors, or spasms. Com- bines well with *Leonurus* and *Pulsatilla*.
	Anemone pul- satilla ⊗ P, △ N, △ Tincture: 1-10 gtt, t.i.d.	Depression with nervousness, diz- ziness, and restlessness. Fearful, sad, con- stant weeping. Frequent excla- mations of sorrow or grief.
	Ferula asafetida ⊗ P Tincture: 5-30 gtt t.i.d.	Depression with nervous irritation, possibly hysteria, headache, dizzi- ness, and flatulence. Tight muscles; may be ticklish, or painful to the touch.

Condition	Medicine	Indications
Depression, situational	*Hypericum per-foratum* Δ P, Δ N, Δ Tincture: 5-30 gtt (or 200-400 mg dried herb), t.i.d. for a short time, then decrease dose to b.i.d. Antidepressant effects require substantial doses (300-400 mg b.i.d.) and take 2-3 weeks to manifest.	For mild to moderate unipolar or situational depression. "Sour stomach + sour attitude" is a common feature.
	Lavandula an-gustifolia Δ P Tincture: 25-50 gtt, t.i.d. Infusion: 1 tsp lavender flowers to 1 cup hot H₂O), steep 10 minutes. 1 cup q.i.d. Note: Essential oil of lavender is toxic and contraindicated in pregnancy.	Indicated for restlessness, poor sleep, foggy, difficult thinking; nervous strain affecting the stomach and intestines.

Condition	Medicine	Indications
Depression, situational, cont'd.	*Lilium tigrinum* Δ Tincture: 1-5 gtt, t.i.d. Use in moderation.	For anxiety with depression. Weeping, muttering under the breath, and fear of solitude are indicators. Often used in menstrual, menopausal or postpartum depression. Overdose can cause saponin toxicity, leading to somnolence, vomiting and purging.
	Selenicereus grandiflorus Δ Tincture: 5-30 gtt, t.i.d.	In depression especially with marked fear, especially useful in menopause, old age, and with heart disease.
	Valeriana officinalis Δ Tincture: 30-50 gtt, t.i.d. (teaspoonful doses may be required for acute use) Fl. Ext.: 5-30 gtt Powdered herb: 2 g (2 #OO capsules), t.i.d.	Indicated for nervous, restless patient, with agitation, a pale face and clammy skin. Imagined worries. Note: protracted use of dried herb can increase anxiety and/or become a nervous depressant. May potentiate effects of CNS depressant drugs.

Condition	Medicine	Indications
Depression, situational, cont'd.	*Anemone pulsatilla* ⊗P, △N, △ Tincture: 1-10 gtt, t.i.d.	Depression with irritability, nervousness, melancholic outlook; tendency to look on the dark side.
	Avena sativa Tincture: 10-50 gtt, t.i.d.	Tonic effects with antidepressant activity. Patient is emotionally reactive to everything.
Depression, obsessive	*Cimicifuga racemosa* ⊗P, △N, △ Tincture: 30-90 gtt, t.i.d. Fl. Ext: 5-30 gtt.	Indicated for hormonal and menopausal depression.
	Ferula asafetida ⊗P Tincture: 5-30 gtt t.i.d.	Depression with nervous irritation, possibly hysteria, headache, dizziness, and flatulence. Tight muscles; may be ticklish, or painful to the touch.

Condition	Medicine	Indications
Depression, obsessive, cont'd.	*Hypericum perforatum* △P, △N, △ Tincture: 5-30 gtt (or 200-400 mg dried herb), t.i.d. for a short time, then decrease dose to b.i.d. Antidepressant effects require substantial doses (300-400 mg b.i.d.) and take 2-3 weeks to manifest.	For mild to moderate unipolar or situational depression. "Sour stomach + sour attitude" is a common feature.
	Albizzia julibrissin Tincture or Fl. Ext: 15-60 gtt, q.i.d.	Indicated for chronic grief.
	Anemone pulsatilla ⊗P, △N, △ Tincture: 1-10 gtt, t.i.d.	Despondency, sadness, depression, tendency to weep.
Grief, despondency	*Strychnos ignatii* △ Best given in 3X attenuation (1:1000).	Melancholic, hysterical, disposition to grieve.

Condition	Medicine	Indications
Grief, despondency, cont'd.	*Lavandula angustifolia* △P Tincture: 25-50 gtt, t.i.d. Infusion: 1 tsp lavender flowers to 1 cup hot H_2O), steep 10 minutes. 1 cup q.i.d. Note: Essential oil of lavender is toxic and contraindicated in pregnancy.	Indicated for restlessness, poor sleep, foggy, difficult thinking; nervous strain affecting the stomach and intestines. Combines well with *Rosmarinus officinalis*, *Cola nitida*, *Avena sativa* in this usage.
	Panax quinquefolium △P Tincture: (cultivated) 10-50 gtt, t.i.d. (wild) 5-10 gtt, t.i.d. Infusion (1Tbsp dry herb to 1 cup hot H_2O): 2-4 oz t.i.d. Powdered herb: 10-60 gr.	Adaptogen, alterative, antispasmodic, cardiotonic, estrogenic, psychotonic, stimulant, stomachic, and tonic (esp. liver). Some authorities advise caution in pregnancy due to mild estrogenic effects.
	Piper methysticum ⊗P Tincture: 5-60 gtt t.i.d. Fl. Ext.: 5-60 gtt, t.i.d. Powdered herb: 14 gr (2 #OO capsules) t.i.d.	For anxiety with muscle tension, bruxism, restless leg syndrome, and pain.

Condition	Medicine	Indications
Grief, despondency, cont'd.	*Rosmarinus officinalis* Δ P Tincture: 25-50 gtt, t.i.d. Infusion (1 oz herb to 1 pint H_2O): 2-4 oz t.i.d. Oil: 3-6 gtt in a binding vehicle, and taken with H_2O.	Indicated in depressive states with general debility and cardiovascular weakness. Combines well with *Avena sativa*, *Cola nitida*, and *Verbena officinalis*.
	Strychnos ignatii Δ Best given in 3X attenuation (1:1000).	Classic melancholic mood; hysterical, disposition to grieve.
	Bacopa monnieri Tincture: 30-40 gtt, t.i.d. Powdered herb: 300 mg q.d.	Mental confusion with anxiety, and nervous exhaustion with agitation. Especially indicated for "cloudy thinking" and poor retention of new information. Contraindicated in hyperthyroidism.
	Cimicifuga racemosa ⊗ P, Δ N, Δ Tincture: 30-90 gtt, t.i.d. Fl. Ext: 5-30 gtt.	Indicated for hormonal source depression and agitation.

Condition	Medicine	Indications
Hyperactivity/ restlessness	*Datura stramo-nium* △ Tincture: 1-5 gtt t.i.d.	Anodyne, anti-cholinergicantich olinergic, anti-spasmodic, antispasmodic, sedative; Use in delirium tremens with rage, in-clined to violence (self or others).
	Matriarca chamomilla △ P Infusion (1Tbsp dry flowers to 1 cup hot H_2O), 2-6 oz PRN.	Caution in preg-nancy if using parts of the plant other than the flowers, due to emmenagogue effects.
	Passiflora in-carnata ⊗ P Tincture: 30-50 gtt up to q.i.d. Fl. Ext.: 30-60 gtt up to q.i.d. Powdered herb: 5-90 gr. up to q.i.d.	Specific for nerv-ous restlessness; sleeplessness with muscle twitching, or cir-cular thinking.
	Piper methysti-cum ⊗ P Tincture: 5-60 gtt t.i.d. Fl. Ext.: 5-60 gtt, t.i.d. Powdered herb: 14 gr (2 #OO capsules) t.i.d.	For anxiety with muscle tension, bruxism, restless leg syndrome, and pain.

Condition	Medicine	Indications
Hyperactivity/ restlessness, cont'd.	*Atropa bella-donna* △ Best given in 3X (1:1000) at-tenuation (homeopathic dose); Tincture can be added to 4 oz. H_2O and ½ tea-spoon doses given t.i.d.: Tincture (leaf) 2-5 gtt Tincture (root) 1 gt.	Indicated in nerv-ous excitement, hysterical presen-tation with delirium, dilated pupils.
	Ferula asafetida ⊗P Tincture: 5-30 gtt t.i.d.	Depression with nervous irritation, possibly hysteria, headache, dizzi-ness, and flatulence. Tight muscles; may be ticklish, or painful to the touch.
Hysteria	*Lobelia inflata* ⊗P, △ Tincture: 10-20 gtt, q.i.d.	Antispasmodic, depressant, re-laxant.; Contraindicated in bradycardia.
	Passiflora in-carnata ⊗P Tincture: 30-50 gtt up to q.i.d. Fl. Ext.: 30-60 gtt up to q.i.d. Powdered herb: 5-90 gr. up to q.i.d.	Specific for nerv-ous restlessness; sleeplessness with muscle twitching, or cir-cular thinking.

187

Condition	Medicine	Indications
Hysteria, cont'd.	*Turnera diffusa* △P Tincture: 10-30 gtt, t.i.d. Fl. Ext.: ½ to 1 dr. Powdered herb: 15-45 gr.	For anxiety neurosis, usually with predominantly sexual factor.
	Veronicastrum virginicum Tincture: 2-5 gtt q.i.d.	Indicated in tenderness and heavy pain in the hepatic region with somnolence, headaches behind eyes, dizziness and depression; liver-associated depression is the key phrase. Combines with *Avena sativa, Cola nitida,* according to indications.

Condition	Medicine	Indications
Hysteria, cont'd.	*Xanthoxylum spp.* ⊗P, ⊗N Tincture (berries): 5-30 gtt a.c., t.i.d. Tincture (bark): 2-20 gtt, a.c., t.i.d. Infusion (1 oz powdered herb to 1 qt hot H_2O, steep for 10 minutes): 1 oz of infusion q2h in acute cases, t.i.d. otherwise. Powdered herb: 5-20 gr, t.i.d.	*Xanthoxylum americanum, carolinianum* or *fraxineum* are essentially the same. Especially indicated in weak, anemic patients.
	Cimicifuga racemosa ⊗P, △N, △ Tincture: 30-90 gtt, t.i.d. Fl. Ext: 5-30 gtt.	Indicated for "gloom and doom" depression, hormonal depression and mood swings.
	Ferula asafetida ⊗P Tincture: 5-30 gtt t.i.d.	Depression with nervous irritation, possibly hysteria, headache, dizziness, and flatulence. Tight muscles; may be ticklish, or painful to the touch.

Condition	Medicine	Indications
Labile emotions	*Anemone pulsatilla* ⊗P, △N, △ Tincture: 1-10 gtt, t.i.d.	First sleep is restless; Nervousness, sadness, depression, tendency to weep.
	Atropa belladonna △ Best given in 3X (1:1000) attenuation (homeopathic dose); Tincture can be added to 4 oz. H₂O and ½ teaspoon doses given t.i.d.: Tincture (leaf) 2-5 gtt Tincture (root) 1 gt.	Indicated in nervous excitement, hysterical presentation with delirium, dilated pupils.
Nightmares and Night Terrors	*Cimicifuga racemosa* ⊗P, △N, △ Tincture: 30-90 gtt, t.i.d. Fl. Ext: 5-30 gtt.	Indicated for hormonal depression and sleep disturbances.
	Strychnos ignatii △ Best given in 3X attenuation (1:1000).	Melancholic, hysterical, disposition to grieve.

Condition	Medicine	Indications
Nightmares and Night Terrors, cont'd.	*Ferula asafetida* ⊗ P Tincture: 5-30 gtt t.i.d.	Depression with nervous irritation, possibly hysteria, headache, dizziness, and flatulence. Tight muscles; may be ticklish, or painful to the touch.
	Ocimum sanctum Tincture: 30-50 gtt t.i.d. Infusion (1 Tbsp dry herb to 1 cup hot H_2O): 2-6 oz t.i.d.	Indicated when the patient is fixated on a specific topic or event. Usually fatigued and mentally foggy. Combines well with *Lavandula*, *Hypericum*, and *Rosmarinus*.
	Scutellaria lateriflora ⊗ P Tincture: 20-60 gtt t.i.d. Infusion (1 Tbsp dry herb to 1 cup hot H_2O): 2-6 oz t.i.d.	Nervousness, twitching, or spasms due to mental or physical exertion. Nervousness without apparent cause. Anger, irritability, angry outbursts.

Condition	Medicine	Indications
Obsessive Thinking	*Angelica sinensis* ⊗P, △ Tincture: 30 gtt, t.i.d. Infusion (1Tbsp dry herb to 1 cup hot H_2O): 2-3 Tbsp, t.i.d. Powdered herb: 30-60 gr.	May trigger spotting in pregnancy. Avoid use in diabetics (increases blood sugar) and in peptic ulcer or acid reflux. Large doses may increase blood pressure and cause photosensitivity.
	Borago officinalis ⊗N Tincture: 10-40 gtt (or infusion of 2 teaspoons dried herb to one cup boiling H_2O, steeped 10-15 min.), t.i.d.	Action: oil from seeds influences prostaglandin synthesis.
	Chamaelirium luteum △ P Tincture: 10-40 gtt (or infusion of 1 oz dried herb to one pint boiling H_2O, steeped 10-15 min.), 1 cup t.i.d.	Indications: sense of heaviness; congestion of pelvic organs; emotional irritability.

Condition	Medicine	Indications
Premenstrual Syndrome (PMS)	*Cimicifuga racemosa* ⊗P, △N, △ Tincture: 30-90 gtt, t.i.d. Fl. Ext: 5-30 gtt.	Indicated for "gloom and doom" depression, hormonal depression.
	Leonurus cardiaca △P Tincture: 10-30 gtt, t.i.d	Mild sedative and anxiolytic. Combine with *Verbena officinalis* for PMS, menstrual or menopausal anxiety. Add *Anemone pulsatilla* for panic attacks.
	Lilium tigrinum △ Tincture: 1-5 gtt, t.i.d. Use in moderation.	For anxiety with depression. Weeping, muttering under the breath, and fear of solitude are indicators. Often used in menstrual, menopausal or post-partum depression. Overdose can cause saponin toxicity, leading to somnolence, vomiting and purging.

Condition	Medicine	Indications
Premenstrual Syndrome (PMS), cont'd.	*Linum usitatis-simum* (flax seed) ⊗P Oil: 1-2 Tbsp q.d. Seeds: 2 Tbsp b.i.d. or t.i.d.	Action: oil from seeds influences prostaglandin synthesis.
	Oenothera bi-ennis Δ (Evening Prim-rose oil) Oil: 1-3 tsp q.d. OO capsules: 2-6 q.d.	Action: oil from seeds influences prostaglandin synthesis; caution in epileptic pa-tients.
	Ribes nigra (black currant) Oil: 1-2 tsp. q.d. Tincture of leaves: 15-30 gtt b.i.d.	Action: oil from seeds influences prostaglandin synthesis.
	Taraxacum officinalis Tincture: 30-50 gtt, t.i.d.	Hepatotonic; helps liver conju-gate and metabolize estro-gen; also a diuretic.
	Trifolium prat-ense ⊗P Tincture: 10-30 gtt, t.i.d. Fl. Ext.: 3-10 gtt.	Alterative, anti-spasmodic, es-trogenic. Contra-indicated in pa-tients with bleed-ing disorders, or taking aspirin / anticoagulants, or within two weeks of surgery, and in renal failure, Graves' disease, and diabetes.

Condition	Medicine	Indications
Premenstrual Syndrome (PMS), cont'd.	*Verbena offici-nalis* △P Tincture: 15-20 gtt, t.i.d.	Premenstrual anxiety; anxiety with nervous tics, tremors, or spasms. Combines well with *Leonurus* and *Pulsatilla*.
	Viburnum pruni-folium ⊗P Tincture: 30-60 gtt up to q.i.d. Fl. Ext.: 15-40 gtt, t.i.d. Infusion (1 oz herb to 1 pint H_2O): 1 cup t.i.d.	A uterine sedative and tonic. *Viburnum opulus* has the same actions but is not contraindicated in pregnancy.
	Vitex agnus castus ⊗P Tincture: 10-30 gtt, t.i.d.	Indicated in PMS based on hyperfolliculinism; also acne and premenstrual herpes eruptions.
	Albizzia julibris-sin Tincture or Fl. Ext: 15-60 gtt, q.i.d.	A powerful mood elevator; useful for anxiety, fears, irritability, insomnia, nightmares, PTSD, moodiness, and disappointment. Do not use in manic patients.

Condition	Medicine	Indications
Premenstrual Syndrome (PMS), cont'd.	*Viburnum prunifolium* ⊗P Tincture: 30-60 gtt up to q.i.d. Fl. Ext.: 15-40 gtt, t.i.d. Infusion (1 oz herb to 1 pint H$_2$O): 1 cup t.i.d.	A uterine sedative and tonic. *Viburnum opulus* has the same actions but is not contraindicated in pregnancy.
	Panax quinque-folium △P Tincture: (cultivated) 10-50 gtt, t.i.d. (wild) 5-10 gtt, t.i.d. Infusion (1Tbsp dry herb to 1 cup hot H$_2$O): 2-4 oz t.i.d. Powdered herb: 10-60 gr.	Adaptogen, alterative, antispasmodic, cardiotonic, estrogenic, psychotonic, stimulant, stomachic, and tonic (esp. liver). Some authorities advise caution in pregnancy due to mild estrogenic effects.
PTSD (post-traumatic stress disorder)	*Valeriana offici-nalis* △ Tincture: 30-50 gtt, t.i.d. (teaspoonful doses may be required for acute use) Fl. Ext.: 5-30 gtt Powdered herb: 2 g (2 #OO capsules), t.i.d.	Indicated for nervous, restless patient, with agitation, a pale face and clammy skin. Note: protracted use of dried herb can increase anxiety and/or become a nervous depressant. May potentiate effects of CNS depressant drugs.

Condition	Medicine	Indications
PTSD (post-traumatic stress disorder), cont'd.	*Avena sativa* Tincture: 10-50 gtt, t.i.d.	Indicated in nervous exhaustion, low libido, insomnia, possibly tremors. Patient is emotionally reactive to everything.
	Ferula asafetida ⊗P Tincture: 5-30 gtt t.i.d.	Depression with nervous irritation, possibly hysteria, headache, dizziness, and flatulence. Tight muscles; may be ticklish, or painful to the touch. A nerve tonic and used in convalescence.
	Humulus lupulus Tincture: 30-50 gtt t.i.d., or infusion (1Tbsp dry herb to 1 cup hot H_2O), 2-4 oz t.i.d.	Mild anxiety, nervous irritability, and wakefulness, possibly with gastric symptoms.
	Bacopa monnieri Tincture: 30-40 gtt, t.i.d. Powdered herb: 300 mg q.d.	Mental confusion with anxiety, and nervous exhaustion with agitation. Especially indicated for "cloudy thinking" and poor retention of new information. Contraindicated in hyperthyroidism.

Condition	Medicine	Indications
Seasonal Affective Disorder (SAD)	*Hypericum per-foratum* Δ P, Δ N, Δ Tincture: 5-30 gtt (or 200-400 mg dried herb), t.i.d. for a short time, then decrease dose to b.i.d. Antidepressant effects require substantial doses (300-400 mg b.i.d.) and take 2-3 weeks to manifest.	For mild to moderate unipolar or situational depression. "Sour stomach + sour attitude" is a common feature. Large doses can create photosensitivity. Use caution in patients undergoing heliotherapy or ultraviolet therapy.
	Melissa offici-nalis Δ P Tincture: 30-50 gtt t.i.d. Infusion (1 Tbsp dry herb to a cup of H_2O): 1 cup q1-2h PRN for pain.	A mild mood elevator. Use in combination with *Hypericum* for SAD. Antidepressant, antispasmodic, and sedative. Insomnia due to nervous strain; gastrointestinal problems. Contraindicated in hypothyroid patients. Lower doses in pregnancy.

Condition	Medicine	Indications
Seasonal Affective Disorder (SAD), cont'd.	*Scutellaria lateriflora* ⊗ P Tincture: 20-60 gtt t.i.d. Infusion (1 Tbsp dry herb to 1 cup hot H_2O): 2-6 oz t.i.d.	Nervousness, twitching, or spasms due to mental or physical exertion. Nervousness without apparent cause. Anger, irritability, angry outbursts.

Bibliography

- Brinker, F., *The Toxicology of Botanical Medicines*, 2nd Edition, Eclectic Medical Publications, 1986
- Ellingwood, F., *American Materia Medica, Therapeutics and Pharmacognosy,* Eclectic Medical Publications, 1994
- Felter, H., *The Eclectic Materia Medica, Pharmacology and Therapeutics*; Eclectic Medical Publications, 1985
- Felter, H., and Lloyd, J.U., *King's American Dispensatory Vol. I and II*, Eclectic Medical Publications, 1983
- Jones, E.G., MD, *Definite Medication*, Therapeutic Pub. Co., 1911
- Jones, E., *Reading The Eye, Pulse, and Tongue For The Indicated Remedy*, Buckeye Naturopathic Press 1989
- Kuts-Cheraux, A.W., *Naturae Medicina and Naturopathic Dispensatory,* American Naturopathic Physicians and Surgeons Assn., 1953
- Sherman, J.A., *The Complete Botanical Prescriber*, self-published, 1979
- Winston, D., *Herbal Therapeutics, Specific Indications For Herbs & Herbal Formulas*, HTRL, Washington, NJ, 2000
- Winston, D., Eclectic Specific Condition Review: Depression, in the *Protocol Journal of Botanical Medicine*, 2(1): 72-73, 1996

13

HOMEOPATHIC MEDICINES

The homeopathic school of medicine is based on the principle that minute doses of substances, which in large doses would cause a set of symptoms in a healthy person, will relieve those symptoms in a sick person. It appears to provoke a response that is diametrically opposite to the behavior of the body that is desired to change[52].

Any substance can be used homeopathically: animal, mineral, chemical, or vegetable (80% of homeopathic medicines are plant-derived). One hundred years ago, 20% of all physicians in urban areas in the USA were homeopaths. Political pressure, on the part of the allopathic medical community and the emerging pharmaceutical giants, resulted in the profession being almost entirely eliminated, with the last medical school teaching homeopathic courses discontinuing its homeopathic curriculum in 1960. Today, Homeopathy is a postgraduate medical specialty with no four-year colleges or hospitals to its name in America, although it continues to have a strong presence in other countries.

Homeopathic medicines, as the reader may know, are made by taking the base substance and grinding or diluting it in a 1:9 ra-

---

[52] This principle is not purely physical in its actions. Psychosocial efforts and counseling make use of it all the time. So-called "reverse psychology", in the common parlance, is an example of the inadvertent use of homeopathic principles to cancel out the pathology.

tio[53] many times. Each time, the mixture is violently agitated (called *succussion*). The first time this is done, the substance is reduced to a 1/10 concentration. This is referred to as a "1X" potency. The next time, it is a 1/100 strength (2X). The next, it is a 1/1000 strength at a 3X potency[54]. By the time a 6X potency is reached, the medicine contains only one one-millionth part of the original substance. Yet this is only the starting point for homeopathic medicines.

When homeopathic doctors were plentiful, they were divided into low-potency prescribers and high-potency advocates. Dissention between the "highs" and "lows" weakened the movement as much as pressure from without. Both ranges of medicines worked. The "lows" tended to be more pathological prescribers, and for this, the lower potencies were better, because they had more of an action on tissue changes. The higher potencies, on the other hand, had more of an impact on the mental/emotional sphere, and the doctors who prescribed them took a different viewpoint. From their perspective, illness derived from psychic imbalances, and to cure physical disease, one had to match the medicine up with the patient's mental state more precisely than with the physical condition.

Why use medicines that are obviously getting weaker as they are refined? The succussion at every level of dilution activates the intrinsic information in the substance and appears to amplify it, even as molecules of the substance are being gradually filtered out. As anyone familiar with *Avogadro's number* will realize, past the 24th dilution, *no* molecules of the original substance are likely to remain. Yet many of the medicinal effects of the base substance, instead of being absent, are *magnified* by this process. This conundrum is what has caused homeopathic medicines to be so hotly debated for so long. No one wants to have to rewrite the physics textbooks, and if Homeopathy is valid, it fol-

---

[53] Another scale is also used, called the *centesimal* scale, which dilutes in a 1:100 ratio.
[54] When the *decimal* scale is used, the suffix used is "X". When the *centesimal* scale is used, the suffix is "C". Therefore, at 1/100 drug strength, a 2X has the same dilution factor as a 1C.

lows that our view of the physical world is badly flawed.

As impossible as it seems, controlled trials have scientifically verified the biological activity of highly diluted or "high potency" homeopathic medicines, even in dilutions with no possible chemical action. Studies using such state-of-the-art technology as thin-layer chromatography and nuclear magnetic resonance have also demonstrated the distinct differences between potentized substances and simply diluted (non-succussed) ones.

The turning point for conventional physicians and scientists to take Homeopathy seriously came with the September 20, 1997 issue of *The Lancet*, which published a meta-analysis of 89 blinded, randomized, placebo-controlled clinical trials. The authors concluded that the clinical effects of homeopathic medicines are not simply the results of placebo. The end result was that homeopathic medicines had 2.45 times greater an effect than placebo.

Why do such impossibly small doses have an effect? It is because of the long-held observation that a sick person is hypersensitive to the specific medicinal substance that matches most closely the total symptom picture of the person. It is assumed that some type of electro-magnetic resonance exists between the medicine and the patient, and that the medicine is acting in an energetic, not chemical, fashion.

Because the homeopathic system bases its selection of the medicine on the ability of the original substance to cause the similar symptoms (in overdose), we have a more-or-less clear path to follow in order to match up the correct medicine with the person. Careful case taking is necessary, therefore, in order to elicit the best information for making that choice of medicine.

This methodology is all the more sensible when one returns to the concept—an inherent element of naturopathic philosophy— that disease symptoms are not just the result of tissue breakdown or the invasion of the body by outside forces, but are the body's reactions to correct those events. Inflammation, pain, fever, exudates, metabolic changes, and other adaptations take place as a *corrective response* to the situation. On the psychological level, reactions and adaptations to emotional trauma or stresses are no different. They are defenses and attempts to heal. The key to healing, in other words, is provided by the symptoms themselves. *How* the person is reacting is an indication as to how to empower the body-mind system to complete the job and re-establish normalcy.

It should now be more clear to the reader who is investigating these concepts for the first time that treating on the basis of the pathological "label"—this syndrome or that neurosis—is super-ficial and ultimately ineffective.

To return to the process of prescribing homeopathic medicines for a person with psychoemotional problems: As expressed ear-lier, the higher potencies (200C, 1M, 10M, 50M) are the most effective for these conditions. The greater the similarity be-tween the collective symptom picture of the patient and the profile of the medicine, the higher the potency can be prescribed with confidence. On the other hand, if there is any doubt, it is a much better policy to first give a mid-range potency (30X or 30C) to assess response. If there are positive changes, then one can follow up with a higher potency later. If not, the case can be re-examined for a different prescription.

If only a few symptoms are present in the patient, a mid-range potency that covers them should be given. It is likely that after the response to the low potency, other symptoms will emerge, which indicate another remedy that can be given in higher po-tency. Often, it is after an initial prescription that a more accurate symptom picture emerges.

With the right medicine, there should be a positive response with any potency. But in emotional illness, the low to mid-range po-tencies are unlikely to provide any deep or lasting relief.

Guide to the Most Indicated Homeopathic Medicines
By Condition

Addiction Withdrawal: alcohol
ALCO, ANT-T, ARS, CALC, CAPS, CAUST, COCC, CON, HEP, LACH, MAG-C, MERC, NUX-V, OP, PETR, PULS, STAPH, STRYCH, SULPH

Addiction Withdrawal: drugs
(See also nightmares)
ARS, AVENA, CANN-I, CANN-S, CIMIC, CON, GELS, HYOS, NUX-M, NUX-V, OP, SEC, STAPH, STRAM, SUMB

Addiction Withdrawal: tobacco
CALAD, DAPHNE, GELS, LOB, PLANT

Adrenal Dysfunction contributing to symptoms:
Adrenal stress: AM-C, CALC, FERR, GRAPH, THYR
Adrenal fatigue: ADREN, ARG-N, ARS, CALC-AR, CHIN, IOD, FERR-I, PHOS, PHOS-AC, SIL, SULPH, VANAD

Alzheimer's Disease:
BELL, BRY, CALC, CANN-I, CARB-V, GLON, LACH, MERC, NAT-M, NUX-M, NUX-V, ONOS, OP, PETR, RHUS-T, SEP, SIL, STRYCH

Anxiety
ACON, ARG-N, ARS, ARS-I, AUR, BELL, BISM, BRY, CACT, CALC, CALC-P, CALC-S, CAMPH, CANN-I, CARB-S, CARB-V, CAUST, CON, DIG, IOD, KALI-AR, KALI-C, KALI-P, KALI-S, LYC, MEZ, NAT-A, NAT-C, NIT-AC, PHOS, PULS, RHUS-T, SEC, SULPH, VERAT

Anxiety, menopausal
ACON, CACT, CAUST, IOD, LIL-T, PULS, SEP, THUJ

Anxiety of pregnancy
ACON, CHAM, CIMIC, COFF,

Agitation due to overwork or exertion
ARN, KALI-P, NUX-V, PHOS, PHOS-AC, STAPH, TUB

Apathy, despondency
AGAR, APIS, BAPT, BRY, CIMIC, CHIN, GELS, HELL, IGN, LACH, LIL-T, NAT-M, NUX-V, OP, PHOS-AC, PHOS, PIC-AC, PULS, SEP, STAPH

Appetite, decreased:
ABIES-N, ANT-C, ARS, BAPT, CARB-AC, CHEL, CHIN, CHIN-AR, FERR, GENT, IGN, IPEC, LYC, NUX-V, PULS, RHUS-T, SEP, SULPH

Appetite, increased:
ABROT, ANAC, CALC, CINA, IOD, LYC, NAT-M, NUX-V, PETR, PHOS, PSOR, SULPH, THYR, URAN-N, ZINC

Asthenia, exhaustion
ANAC, CALC-P, CARB-V, CHIN, COCC, HELON, KALI-P, PH-AC, PHOS, PSOR, SEL, SIL, STAPH, SUL-AC, ZINC

Attention Deficit Disorder
ANAC, APIS, ARG-N, ARS, AUR, BAR-C, BELL, CALC, CALC-P, CANTH, CARBO-V, CARB-S, CAUST, CHAM, CIC, CINA, CUPR, HEP, HYOS, KALI-C, LAC-C, LACH, LYC, LYSS, MED, MERC, NAT-M, NIT-AC, NUX-V, PETROL, PHOS, PLAT, PULS, SEP, SIL, STRAM, SULPH, TUB, VERAT, ZINC

Bipolar Affective Disorder: manic phase
ACON, ANAC, ARG-N, ARS, AUR, BAPT, BELL, PHOS, CALC-P, CAMPH, CANN-I, CARC, CIMIC, COLOC, CUPR, FERR, LACH, LYC, MED, MERC, RHUS-T, SEC, SEP, SIL, STAPH, STRAM, SULPH, TARENT, VERAT, ZINC

Bipolar Affective Disorder: depressive phase
ACON, AUR, CALC, CARBO-AN, CARB-S, CARC, CAUST, CHIN, CIMIC, FERR, FERR-I, GRAPH, HELL, IOD, IGN, KALI-BR, KALI-P, LAC-C, LACH, MERC, MEZ, MURX, NAT-C, NAT-M, NAT-S, RHUS-T, PSOR, PULS, SEP, SULPH, THUJ, VERAT, ZINC

Confusion
ALCO, AETH, ANAC, ARG-N, AUR-M, CALC, CANN-I, CAPS, CON, DIG, GELS, KALI-BR, KALI-P, LYC, NAT-C, NUX-M, NUX-V, OP, PHOS-AC, PHOS, PIC-AC, PLB, RHUS-T, SEP, SIL, ZINC

Delusions
ACON, AGAR, ARS, BELL, CANN-I, CANN-S, DROS, HYOS, KALI-BR, LACH, PLAT, STRAM

Depressive Anxiety
AMBRA, ANAC, ARG-N, ASAR, BELL, BOR, CHAM, CIMIC, COFF, FERR, GELS, HYOS, IGN, KALI-BR, KALI-P, LACH, LIL-T, MAG-C, MED, NAT-C, NUX-V, PHOS, PSOR, PULS, SEC, SEP, SIL, STRAM, SUMB, VAL, ZINC

Depression, endogenous
AGN, ALUM, ANAC, ARS, AUR, CIMIC, CHIN, CON, CYCL, GRAPH, HELON, HYDR, IGN, INDIG, LIL-T, LYC, MUX-V, PHOS-AC, PHOS, PLAT, PLB, PSOR, PULS, SEP, STANN, STAPH, TUB

Depression, obsessive
AGN, ANT-C, AUR, CYCL, GRAPH, IGN, INDIG, LACH, LIL-T, NAT-M, PULS, SELEN, SEP, SIL, STANN, STAPH,

Grief, despondency
AUR, CAUST, IGN, LACH, NAT-M, PHAC, PHOS, SEP

Hyperactivity/restlessness
ACON, ANAC, ARG-N, ARS, BAPT, CALC-P, CAMPH, CARC, CIMIC, COLOC, CUPR, FERR, LYC, MED, MERC, RHUS-T, SEC, SEP, SIL, STAPH, STRAM, SULPH, TARENT, ZINC

Hysteria
ACON, AMBRA, ASAF, CIMIC, COFF, CROC, GELS, IGN, KALI-P, LIL-T, MOSCH, NAT-M, NUX-V, PLAT, PULS, SEP, SUMB, TARENT-H, ZINC-V

Labile Emotions
ACON, ALUM, AMBRA, ASAF, CIMIC, CROC, IGN, MOSCH, NUX-M, PLAT, PULS, VAL, ZINC-V

Nightmares and Night Terrors
ACON, AUR-BR, CALC, CHAM, CIC, CINA, KALI-BR, KALI-P, SCUT, STRAM, TUB, VERAT

Obsessive Thinking
ARS, HYOS, IGN, MED, NUX-V, PULS, STAPH, THUJ

Premenstrual Syndrome (PMS)
CIM, COLOC, FOLLIC, LACH, OOPH, PLAT, SEP, XAN

PTSD (post-traumatic stress disorder)
ARN, ARS, STRAM

Seasonal Affective Disorder (SAD)
ARG-N, BROM, CALC, CAUST, MED, PHOS, PULS, STRAM

Suicidal Ideations
ARS, AUR, CALC, CIMIC, IGN, LIL-T, LYC, SEP, SULPH

Tourette's Syndrome:
HYOS, SCOP, STRAM, TARENT, VERAT

Bibliography

Microdose research:
- Benveniste, J., Davenas, E., Ducot, B., et al., *L'agitation de Solutions Hautement Diluees n'induit pas d'activite Biologique Specifique*, C.R. Acad. Sci. Paris, 1991, 312:461
- Ferley, J.P., Zmirou, D., D'Admehar, D., et al., A Controlled *Evaluation of a Homoeopathic Preparation in the Treatment of Influenza-like Syndrome*, British Journal of Clinical Pharmacology, March 1989, 27:329-35.
- Reilly, D., Taylor, M., McSharry, C., et al., *Is Homoeopathy a Placebo Response? Controlled Trial of Homoeopathic Potency, with Pollen in Hayfever as Model*, The Lancet, October 18, 1986, 881-86
- Stebbing, A.R.D., *Hormesis: The Stimulation of Growth by Low Levels of Inhibitors*, Science of the Total Environment, 1982, 22: 213-34.
- Ullman, D., *Homeopathic Medicine is Nanopharmacology*, The Institute of Science in Society 2006 web site (http://www.i-sis.org.uk/nanopharmacology.php)

Homeopathic methodology:
- Boericke, W., and Boericke, O., *Pocket Manual of Homeopathic Materia Medica With Repertory*, Boericke & Runyon, 1927
- Chappell, P.; *Emotional Healing with Homeopathy*, Element Books Ltd., 1994
- Clarke, J.H., *A Dictionary of Practical Materia Medica*, Homeopathic Publishing Co., London 1902
- Dewey, W., *Practical Homeopathic Therapeutics*, Boericke & Tafel, 1901
- Gallavardin, L., *How To Cure Alcoholism The Non-toxic Homoeopathic Way*, East-West Arts, 1976
- Kent, J., *Repertory of the Homeopathic Materia Medica*, 5th Edition, Ehrhart & Karl, 1945
- Murphy, R., *Homeopathic Medical Repertory*; Hahnemann Academy of North America, 1993
- Smith, T., *Homeopathic Medicine for Mental Health*, Healing Arts Press 1989

14

INTEGRATED METHODOLOGY

This chapter will introduce the reader to the concept of building a treatment plan based on the initial assessment of the patient or client and choosing modalities that are available, to begin having an impact on the person's problems. Whether the clinician applies manual therapies, body-based psychotherapy, acupuncture, or phototherapy in the office, or refers out for them; whether botanical medicines and/or nutraceuticals are dispensed or not, it is helpful to have more than one mode of action for therapeutic changes to happen—even if as simple as having lavender oil (for example) infused in the consulting room.

The psycho-emotional-neuro-endocrine (PENE) system will be impacted at different points to different degrees. Therefore, people with problems need different things, and at different times. Some may improve dramatically from simply inhaling a fragrance; some may need desperately to be touched, and no oral medications of any kind will be enough to launch a healing of the emotions. A manual therapy will be necessary for these individuals.

As often happens, more critical and individualizing information emerges as one comes to know the person better and observes his or her response to the therapies already used. Then a more precise homeopathic prescription can be arrived at, and this very often is the turning point in the therapy.

The Holistic Perspective

Mental and emotional problems of the patient can either initiate or maintain somatic problems. It has long been the holistic practitioner's credo that the whole person must be treated. How is this done? Some treat with somatic therapies combined with more-or-less standard psychotherapy. Others may take the view that correcting nutritional deficiencies will usually correct the psychological imbalances, and they are often (but not always) right.

Asian medicine has traditionally looked at psychological problems as another manifestation of energy imbalances, and assumes that correction of the imbalance will result in a clearing of the mental symptoms. Again, this often happens. Sometimes it does not.

Chiropractors have long used the equilateral triangle to represent the human condition, the three sides representing the chemical, psychological, and structural nature of mankind. Anything that disturbs one changes the other two; this is why Chiropractic has always asserted the effect of neuromusculoskeletal treatment on organic function. Fewer people seek out Chiropractors for alleviation of their psychological problems, but techniques adopted by them have been startlingly effective in neutralizing emotional trauma by balancing specific muscles.

Drawing from Applied Kinesiology, Touch For Health is a system embraced by non-chiropractor body workers and achieves the same effects through the same mechanisms and much the same methodology. As previously detailed, balancing the meridian system of acupuncture with muscle challenges and point contacts creates a therapy used by many massage therapists to effectively address issues with mental clarity and mood disorders.

Homeopaths have a schema similar to the chiropractic triangle: they have always looked at people as mental beings wrapped in an emotional layer, and wrapped again in an outer physical layer. In other words, the mental sphere is the deepest and the physical is the most superficial. The "center of gravity", so to speak, of a person's illness may be on one of the three levels, but it will often affect the other two. A problem that has purely physical symptoms will be relatively easy to cure by this thinking; emotional problems are more challenging because they are deeper, and problems that affect the rational ability (schizophrenia, etc.) are the deepest pathologies and are the hardest to cure.

The "classical" (Kentian) homeopath, recognizing that the direction of cure is always outward, seeks above all to find the medicine that corrects the patient on the deepest level. In this way, the healing extends its way toward the physical symptoms that likely brought the patient in for treatment in the first place. The patient on the mend typically feels better subjectively, even while the superficial symptoms are worsened during the treatment process. It is well known that such a sign has a good prognosis.

So far this theory is sound. However, the *method* by which one discovers the medicinal agent that will work this magic is crucial. If it is not effective, the medicine will not be. Homeopaths have traditionally worked out the correct prescription by extensive talking with the patient. The combination of insightful case taking, knowledge of materia medica, and cross-referencing the many medicines that may be indicated is essential. This is a time-consuming art that requires much skill and experience to be consistently successful. But it is dependent also on one other factor: that of the patient.

When the path to gaining the necessary information to treat correctly and effectively involves *talking*, problems enter the picture. Unfortunately, patients are often not honest. Information may be misleading or simply untruthful. Someone who is deliberately misleading the practitioner still has a real problem; the information used to arrive at the prescription, though, will be partially useless. Even the emotions and issues felt most strongly and sincerely by the patient may in fact not be the most crucial ones to treat. A buried trauma, for example, may be much more vital than one that is conscious and current.

For this reason, the author (CPN) has undertaken to make more efficient the process of selecting the correct medicine *for each phase of treatment* (this last phrase is important). Less experienced practitioners naturally have some "hits" and some "misses". Using Hering's Law of Cure to understand the process, one quickly realizes why even experienced prescribers have more "misses" than they would like. There are several reasons.

> First: The constitutional medicine often fails to cure the pathology, even though it is indicated for the *patient*.

> Second: The patient does not necessarily need the ultimate *simillimum* for his or her deepest pathology at the *start* of treatment.

> Third: What works to clear the current imbalances is not what is needed for the older, more buried problems.

> Fourth: The core issues should be treated, according to the Hering's Law of Cure, *in reverse order of appearance*. The immediate issues should have precedence; once stabilized, the next oldest problems should be addressed, etc., working back all the way to the birth trauma, if necessary. This is what is meant by the aforementioned phrase "each phase of treatment".

> Fifth: There will be an incomplete response to even a perfectly indicated medicine if no detoxification accompanies the treatment.

This last statement will surprise some. It appears, though, that when a specific emotional problem is released in therapy, a specific toxin is also released physically. As mentioned in Chapter One, research has suggested that toxins in the body will couple with specific sets of psychoemotional issues. It appears that the intangible information of those issues become somehow trapped in the toxins. This recalls the traditional Oriental medical concept of emotions residing in certain organs—anger in the liver, for example. Note that a pertinent toxin need not be an environmental chemical or noxious pollutant. It could be a residual toxin from a killed microorganism remaining from an old infection. It could be one arising internally from sensitivity to a particular food. The patient may not recover completely while still eating, and reacting to, a food intolerance.

Naturopaths have always observed that during fasting and other detoxification procedures, patients have realizations, revisit old issues, and experience emotional releases. Body workers regularly see spontaneous emotional releases during massage sessions. This is nothing new. But the prescribing of homeopathic medicines has traditionally left out this important factor of *detoxification*, other than to recommend a simple diet, etc. When a homeopathic medicine works as it should, the patient experiences an outpouring of emotion and the doctor is pleased. What that doctor fails to realize is that on those occasions when the medicine does not produce this effect, it likely represents *a toxin that is "stuck"*.

It should also be mentioned that simply trying to "cancel out" the problem with the indicated homeopathic remedy, as many strive for, would not result in lasting improvement, as evidenced in the need for periodic "booster" doses. The patient must willfully and consciously acknowledge and process the release of the toxic psychoemotional state, with the help of the practitioner. Without this activation of the important neurological feedback loop, discussed shortly, complete release is not likely.

The author's (CPN) "Neo-eclectic" orientation extends to the treatment of the psychoemotional case by always attempting, as with physical complaints, to elicit *signs* from the body that the treatment is a correct one. As a certain appearance of the tongue will not only indicate a particular disease but also its corrective medicine, physical signs can indicate the key mental and emotional issues of the patient as well. In addition, the accurate prescription can be found by *objective testing*, not simply matching up the theoretically best remedy with the symptom picture, as classical homeopaths are wont to do.

This approach, while dramatically successful, is also notable for its ability to raise the practitioner with less experience to a level more consistent with the skilled veteran. A sick person should not have to find a doctor with decades of practice—or a specialist—in order to be effectively helped. This work you are reading strives, above all else, to increase the efficiency, efficacy, and success of the different practitioners using it.

Diagnostic Indicators and Objective Evaluation

Many well-known body reflexes are engaged when a person encounters an emotional "hot spot". Changes in muscle tone, pulse, respiration, skin temperature, eye movement, electrical conductivity at acupuncture points, etc., all occur during the stressful event or situation; they also occur when recalling it. It is this fortunate fact that enables us to use these reflexes to measure the patient's reactions to specific topics and issues. Thus can we individualize the case so that a precise therapy can be found.

There are several methods to objectively find a patient's most pertinent emotional issues. One is by muscle testing.

Muscle Testing

In Applied Kinesiology, originated by George Goodheart, D.C., an assessment is made of the pectoral muscle while the patient is lying supine. Emotional factors are tested by having the patient focus on pressing problems, or mentally recall old disturbing events from the past, while the examiner re-tests the muscle strength. A pertinent topic is one that weakens the muscle. It is also observed that rapid eye movements (REMs) are initiated when the patient comes to a strongly disturbing issue.

Then a light contact is made with the examiner's fingers to neurovascular points on the patient's forehead. Contact is maintained until the examiner feels a slight pulsation beneath the skin. Immediate re-testing of the pectoral muscle will reveal restored strength to the muscle if the emotional recall has been effective in neutralizing the old trauma. The emotional neurovascular receptor points are on the frontal eminence, and more will be said about this shortly.

Electrodermal Screening

Electrodermal Screening (EDS), an outgrowth of acupuncture, involves taking electrical conductance measurements on test points located on the surface of the skin. It can be used to objectively measure responses to both emotional stress and also indicate the effective medication to relieve it[55]. Research in Germany and other countries has shown a relationship between specific points and specific internal tissues and processes, which have profound implications for psychoemotional treatment.

The testing process uses a specially calibrated galvanic skin resistance test meter, with a scale of 0-100. 50-65 is considered the normal physiological range. Elevated values at the points generally reflect inflammation or hyperactivity of the structures related to the test point. Low values indicate hypofunction or organic fatigue. A fall from a higher value to a lower one is indicative of degeneration or active infection.

Measurements are made by pressing a probe with a brass tip against the test point and watching for the change in conductivity on the meter.

---

[55] An appropriate therapeutic substance, introduced into the subject's field during the test, will cause normalized readings of a test point that previously measured in an abnormal range. This is called a biological compatibility evaluation, or BCE.

The measurement point (MP) for the Lamina Tecti, for example, will show increased conductivity in cases of strong depression and low readings or indicator drops in cases of psychosis. The lamina tecti is considered the "physiological location of the emotional life", according to Dr. Reinhold Voll.

MP. Limbic System will reveal mania and violent behavior patterns with elevated readings; low readings occur with repressed emotions.

MP. Cerebrum and MP. Hippocampus will show high values in rapid thoughts, low values in poor memory.

MP. Brain Stem reveals overactive mind and sleep disturbances with high values; low values indicate reduced alertness and hypersomnia.

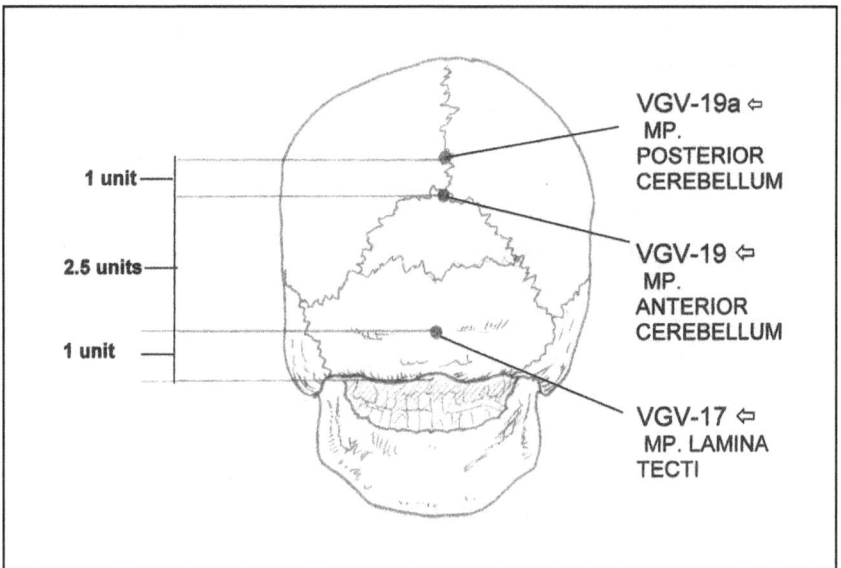

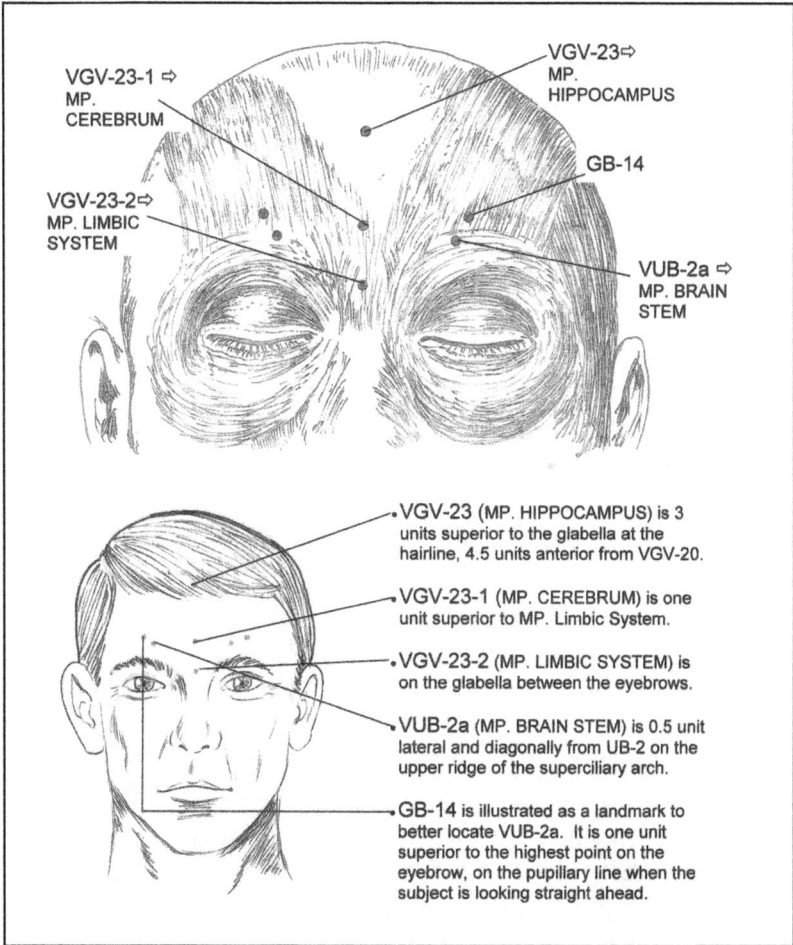

VGV-23-1 ⇨
MP.
CEREBRUM

VGV-23-2⇨
MP. LIMBIC
SYSTEM

VGV-23⇨
MP.
HIPPOCAMPUS

GB-14

VUB-2a ⇨
MP. BRAIN
STEM

• VGV-23 (MP. HIPPOCAMPUS) is 3 units superior to the glabella at the hairline, 4.5 units anterior from VGV-20.

• VGV-23-1 (MP. CEREBRUM) is one unit superior to MP. Limbic System.

• VGV-23-2 (MP. LIMBIC SYSTEM) is on the glabella between the eyebrows.

• VUB-2a (MP. BRAIN STEM) is 0.5 unit lateral and diagonally from UB-2 on the upper ridge of the superciliary arch.

• GB-14 is illustrated as a landmark to better locate VUB-2a. It is one unit superior to the highest point on the eyebrow, on the pupillary line when the subject is looking straight ahead.

All the aforementioned points can be used to test the patient's response to a medicine, depending on the set of problems you are trying to target. A medicine for depression, for example, can be adequately tested by bringing it in contact with the subject and then re-measuring the MP. Lamina Tecti. A corrective medicine will balance the meter at 50 on a scale of 0-100.

One will notice that the measurement point for the brain stem is close to the classical acupuncture point GB-14. The emotional neurovascular receptor points mentioned in the previous discussion of Applied Kinesiology have a strong correlation with both these points, as the fingers will typically cover both points when making the contact. AK practitioners will locate these neurovascular points in roughly the position of the acupuncture point GB-14. It is the author's belief that the effects seen in Applied Kinesiology are due to the correspondence of the contact points with the brain stem. The GB-14 point has no such strong neurological association.

Layers of Emotional Disharmony

It will be found that even when core factors (such as residual effects from a childhood trauma) are discovered in the interview process, the indicated medicine often fails to produce a clearing of the patient's symptoms. This is usually due to a more recent problem blocking the reaction to the clearly indicated medicine. The effective treatment for the current layer will not address the old trauma, and vice-versa. One must work backwards to sequentially neutralize the factors that over the years have collectively resulted in the psychoemotional state of the patient.

Here is an example:

A 22 year-old woman consults her physician for insomnia. She has been unable to sleep well for over six months, taking three or more hours a night to finally go to sleep. She then awakens at some point during the night with a start, and again she has trouble going back to sleep. In the initial interview, no obvious signs of anxiety are apparent. There are no other outstanding health problems that look responsible. A 24-hour cortisol study is ordered and comes back normal. A treatment plan is then put together using protocols from this book:

- Acupuncture treatments once weekly for four weeks, sedating points X-1, Ht-7, PC-6, Sp-6/9 and tonifying points Sp-1, St-45.

- During the acupuncture treatments, the patient under-goes chromotherapy: she is bathed in indigo-filtered light for the duration of the session.
- She is given a prescription of *Passiflora incarnata* tincture, 40 drops twice daily, in the afternoon and 30 minutes before bed.
- She is counseled regarding diet, cautioned about caffeine and sweets in the evening, advised to not go to bed hungry, and told to emphasize foods high in tryptophan: nuts, eggs, meat, fish, dairy. Also, foods high in serotonin (e.g., bananas, pineapples, walnuts, and whole-wheat toast) are recommended.

She reports some mild improvement in her sleeplessness over the course of the next few weeks. When she returns for a re-evaluation, she complains of having a headache. When asked if she gets them often, she replies that being around cigarette smoke is responsible. She says she feels better overall, having more energy during the day and better cognitive abilities at work. She admits that although her sleep is better on the regimen, she is still having a difficult time. She feels calmer in general and takes less time to fall asleep, but she still wakes up at night. She says she feels somewhat depressed about it. When asked to review the history of the problem, she states that she had this problem once before, and she took sleeping pills for some time. In investigating that, the physician finds that this first occurred after her mother died. It has been several years since then.

Now there are two individualizing symptoms to better delineate the case, which both point to the same homeopathic medicine. A causative factor of grief from the death of a parent gives us two major remedies, Causticum (CAUST) and Ignatia (IGN), and a few minor ones (Ars, Calc, Nit-ac, Nux-v, Plat, Staph). Of these, only one has the characteristic indication of headache from smelling tobacco: Ignatia. With a few confirmatory questions, the choice of medicine is obvious. With a book of materia medica in hand, the physician asks: "Do you ever have any strange sensations in your throat?" She confirms that she expe-

221

riences a tightness in her throat often and has a tendency to choke. Bowel movements? She admits to having difficult movements with sharp pains—another Ignatia indication. What has this to do with her sleep problem? Nothing and everything.

The patient is given three doses of Ignatia 200C, taken once daily for three days. When she returns three weeks later, her insomnia is, in her words, "90% better." With an additional dose or two in the future—and only if her sleep gets worse again—she will reach a 100% resolution. As beneficial as the acupuncture and chromotherapy was, as chemically reliable as the botanical medicine is, and as rational as the dietary guidelines are, there was unfinished business in this young woman's psyche. Until her grief was neutralized, the old sleep problem would keep returning despite (seemingly) well-indicated treatments.

Another example:

Our second case is a 32 year-old male office worker. He consults his physician for help with his low back pain, allergies and asthma. He also has some sleep problems; in his case, he is troubled by bad dreams of someone coming into his bedroom while he is sleeping. He also complains of temporomandibular joint syndrome. He wears a mouth splint at night and wonders if it is contributing to his sleep disturbance. He has sought aid because he believes in "mind-body medicine" and wants a holistic approach.

He is clearly an introverted, shy person on observation. He makes fleeting eye contact and presents a mildly anxious mood. He slumps in his chair during the interview. When examined, he has a moderate wheeze bilaterally and a poor breath hold at 30 seconds, and it is obvious that his current medications do not completely control his asthma. There is obvious nasal congestion, and he says his allergy symptoms are year-round. Examination of his back shows a very tight and tender Quadratus lumborum, and tight hamstrings on straight leg raise test, but no sign of discopathy. He has decreased range of motion in the cervical spine also, in left and right rotation, side bending, and extension.

Initial treatment plan:
- Bowen Therapy: low back, upper back, neck and chest protocols; once weekly for 4-5 weeks.
- He is given specific exercises to stretch and strengthen his spinal muscles.
- He is told to rub a few drops of essential oil of *Eucalyptus globulis* on the soles of his feet before bed, cautioned to not inhale the vapors directly so as to not aggravate his asthma.
- He is given a prescription of tincture of *Lobelia inflata*, 10 drops three times daily. If his breathing is improved over the next few weeks, he may be able to decrease his regular medications.

One can see that this case is begun by working simultaneously on the musculoskeletal sources of pain while also helping free the muscular aspect of respiration. The choice of *Lobelia inflata* is likewise for more than one condition. Its use in bronchospasm, of course, is legendary; but if one looks at other indications for it, tight cervical and TMJ muscles respond to it, as well as anxiety. Eucalyptus oil has mucolytic properties as well as bronchodilating capability, and so it can act on the sinuses as well as the bronchi. In this case, absorption through the skin is a safer way of using this type of olfactory therapy.

In three weeks, his back is better from the Bowen Therapy sessions. He says his breathing is less labored, but his sinuses are bothering him terribly. He is given a treatment on local acupuncture points for the sinuses and allergy symptoms and asked to continue the regimen.

The astute observer will recognize that the well-known positive response seen in natural medicine (Hering's Law) is in effect: Healing is taking place in the correct direction—from deep to superficial and primary organ to secondary organ. As the lungs improve, the sinuses temporarily worsen.

223

In one more week, he returns and reports that while his upper respiratory symptoms are not as bad as they were, he has had an emotional crisis. Work-related issues cause him much irritation and, he says, his physical symptoms always get worse under such conditions. He says it has always been this way. He states that he "bottles it up and it makes me sick", and "I can't have a good cry and let it out."

The physician must then focus on his emotional life this session and ask about events, phobias, and aspects of his personality he sees as problematic. He says he is claustrophobic, but it is not a big problem. There have been a few events that have impacted him. When asked about the time line of his emotional trauma, he states that his early childhood was uneventful, but then he attended Catholic school after third grade, and he found this up-setting on almost a daily basis. The nuns who were his teachers were strict, menacing, and verbally abusive. As a quiet and pas-sive individual, he was apparently a target for this behavior often, and he was criticized in front of the other students and made an object of mockery. He thinks his shyness may stem from this.

His next significant emotional trauma was a major love disap-pointment when he was 22.

By consulting the Homeopathic repertory in Chapter Fifteen, it is apparent that a prominent medicine keeps appearing in all the rubrics (the capitalized abbreviations are the strongest indicated medicines for that symptom, italicized ones of secondary strength, and the lower case group the weakest).

His negative school experience as a causative factor:

Reproached, cen-sured with severe language (See **Humiliation**)	**NAT-M**	*Ign, Med, Staph, Stram*	Bell, Carc, Coloc, Nux-v, Ph-ac, Plat, Tarent

Next: disappointed love as a causative factor

Love, disappointed	AUR, HYOS, **NAT-M**, PH-AC, STAPH	*Bell, Caust, Lach*	Kali-c, Nux-v, Phos, Sep, Sulph, Tarent

When asked about the relationship and why it ended, he related that she betrayed him by both leaving him for another and also by revealing confidential personal information to that person.

Betrayal	AUR, LYC, **NAT-M**	*Nux-v*	Ign, Lach, Merc, Ph-ac, Puls, Sep

"I can't have a good cry and let it out"

Grief, but cannot cry	**NAT-M**	*Ign*	Carc, Nux-v, Puls

And the fear of someone breaking into his house:

Intruders will invade home	ARS, **NAT-M**	*Arg-n, Bell, Ign, Lach, Lyc, Merc, Phos*	Anac, Aur, Sil, Sulph

There is a clear standout remedy in Natrum muriaticum (NAT-M). We have rendered it in bold face to call the reader's attention to it. The experienced homeopath knows that one of the conditions it is indicated for is claustrophobia, which the patient admits having. It is also used frequently for allergy symptoms and sinusitis, as well as asthma when the indications are right.

Boericke's Pocket Manual of Materia Medica[56] confirms many of the other symptoms:

- "Pain in back, *with desire for some firm support*" (recall his habit of slumping with his back pressed against the chair back).
- "Painful contraction of hamstrings." (recall the poor straight leg raise)
- "Psychic causes of disease; ill effects of grief, fright, anger, etc."

In this case, the medicine for the current symptoms, past emotional trauma, and the far distant emotional trauma is the same. There is such a consistently strong indication for this remedy, the physician gives it in a single dose of 1M potency and schedule a follow-up in four weeks.

At the next consult, the young man is visibly different. His posture in the chair is more erect, his gaze more steady, and his mood improved. He has had no more bad dreams and is sleeping well. His lung expansion and breath hold is increased. There is no wheezing heard on auscultation. He is told it is now safe to experiment with reducing his use of allopathic bronchodilators and is given a supply of *Lobelia* tincture to use as needed if he gets short of breath. Follow-up consult will be scheduled again at four weeks.

One can see how the multiple modalities of acupuncture, olfactory therapy, manual therapy, botanical medicines, chromotherapy, etc., can be used to stabilize a patient and begin improvement, and often nothing more need be done. This is not to suggest that these modalities are never effective in themselves. We are using these illustrations to spotlight those instances where a person's pathology is so deep that well-indicated measures produce only partial relief. In such cases, we recommend a thorough taking of the case and using the indicated homeopathic

---

[56] Published by Boericke & Runyon, Philadelphia 1927

medicine for each sequential stage of the healing that needs to take place.

Here is the best case to illustrate this:

A 40 year-old woman consults for depression. She also has chronic sinus problems, which she treats symptomatically. Her sinuses have been better lately, but the depression is worsening, and she does not want to go on antidepressants. She is tearful as she talks about it. She has crying spells at odd times throughout the day. She says she feels hopeless.

As the interview progresses, she reveals that she is in doubts about her marriage, and that she has lost affection for her husband in recent years. This apparently has taken place since the birth of their son. She has no sex drive. Her menses have always been irregular, and she thinks she is now perimenopausal. You ask about the birth of her child and she reports having had some difficulty with the birth. An episiotomy was performed. There were no residual problems except that she was told her uterus is prolapsed. This puts pressure on her bladder and she urinates frequently as a result.

The physician begins to build up a time-line picture. She has a family history of cancer, and she is very much afraid of contracting it herself. She had pneumonia at the age of four. No emotional problems growing up, but her father, with whom she was very close, died when she was in college. This was the first time she experienced depression. "I have never gotten over it," she tells you. With additional probing, it is revealed that she was sexually abused in her adolescence by a neighbor. There is more weeping at the revelation of this event.

In order to better visualize these events, they are arranged in a time-line diagram:

Pre-birth: Predisposition to cancer
Birth: Mother said it was smooth and uneventful.
Early years: Peaceful, happy home. Parents had good relationship.
Age 4: Pneumonia
Age 7-present: Allergies and sinus infections
Age 12: Sexual violation
Age 13-present: Irregular menses
Age 22: Father's death; persistent grief and sense of loss
Age 35: Birth of son and episiotomy
Age 36-present: Progressive loss of libido, increase of depression

You ask if there is anything that makes her feel better emotionally. She replies, "Dancing…but we have no opportunity to do it these days."

Using the rubrics that are the strongest-felt *at present,* consulting the repertory produces the following:

First, her mood and affect are taken into consideration.

Crying, tearful mood	APIS, CALC, CALC-S, CARB-S, CAUST, GRAPH, IGN, KALI-BR, LAC-C, LYC, NAT-M, PLAT, PULS, RHUS-T, SEP, SULPH, VERAT	*Acon, Alum, Am-m, Arg-n, Aur, Bell, Calc-p, Carb-v, Cham, Chel, Chin-s, Ci-mic, Coff, Con, Cupr, Dig, Ferr, Hell, Hep, Iod, Kali-bi, Kali-c, Kali-p, Lil-t, Mag-m, Mang, Med, Nat-p, Nat-s, Nit-ac, Nux-m, Ph-ac, Phos, Spong, Staph, Sul-ac, Tarent, Viol-o*	Am-c, Ars, Bar-c, Bor, Cann-i, Colch, Dulc, Gels, Glon, Hura, Lith, Mosch, Nat-c, Psor, Sabin, Sec, Spig, Stram, Thuj, Zinc

There are quite a few possibilities. The medicines, again, are listed as Grade 3 (very strongly indicated for the symptom; all capitals), Grade 2 (strongly indicated; italics), and Grade 1 (indicated; lower case).

With the addition of her depression with hopelessness, the following medicines are indicated:

Depression, despair, hopeless	ARS, AUR, CALC, IGN, LYC, SEP, SULPH	*Anac, Arg-n, Cann-i, Caust, Chin, Lach, Med, Merc, Nat-m, Nit-ac, Nux-v, Puls, Staph, Stram*

Next, one must add the distinguishing characteristic of the problem being helped or ameliorated by something (in this case, dancing). The three medicines that appear prominently in the above symptoms (IGN, LYC, SEP) are now reduced to one Grade 3 remedy (SEP) and one Grade 2 (*Ign*).

Dancing, better from	**SEP**, TARENT (esp. break dancing)	***Ign***, *Nat-m*	Carc, Caust, Sil

Adding the important deviations from the normal state, loss of sex drive and loss of affection for her husband, we see SEP coming to the forefront. This is the most likely medicine to start.

Low or absent sex drive	CAUST, GRAPH, LYC	*Agn, Bar-c, Ferr, Helon, Hep, Mag-c, Nat-m, Ph-ac, Psor, Rhod,* ***Sep***	Alum, Ambr, Bell, Bor, Cann-s, Carb-an, Ferr-p, Kali-i, Mur-ac, Phos, Sil, Sulph
Indifference to loved ones, relatives	PHOS, **SEP**	*Acon, Ars, Plat*	Bell, Carc, Merc

A review of the *materia medica* will show that Sepia (SEP) has within its indications sinus inflammation, uterine prolapse, and irregular menses—all physical symptoms that the patient has as well. A prescription of Sepia 200C is given. The patient takes two doses and is scheduled to return in three weeks.

On her return, she reports that her mood has improved. There have been drastically fewer crying spells, less hopelessness, but still no libido. She was feeling happier in the first two weeks, but in the last few days she is becoming more depressed again. The Sepia 200C is repeated.

At a one-month follow-up visit, she is more concerned with a runny nose than her emotions. She is told that any eliminative symptoms such as these are good, and that it will pass.

Four weeks later, she comes to the office looking very sad. She does not feel the "hopelessness and black depression" as before. She admits to feeling angry "all the time". Consulting the time-line chart, one can see that the case is working itself backwards in time, and the more recent symptoms have cleared, opening the door to older trauma.

In this case, the episiotomy she received at her son's birth was not purely a physical trauma. The body registers such an event as a violent invasion, even if the anesthetized patient does not. The body does not differentiate between actual rape and surgical "rape", and the memory of such is stored in the body-mind system. Such an invasion has much more impact on a patient who was actually sexually assaulted in her youth. Procedures such as Caesarian sections, etc., register with the body as a violent invasion, despite the loss of sensation and the happiness associated with the birth event. Sexual dysfunction can follow much later. So now both the obstetrical and the earlier sexual trauma are coming to the surface for the patient. The emotions she feels at this point constitute a new symptom picture to prescribe for.

Remembering the incident of sexual abuse, she related the dual feelings of terror at the time of the event, then shame and anger, as she kept quiet about it. Looking at traumatic causative factors, we see:

Rape	STRAM		Aur, Sil, Tarent

Stramonium (STRAM) is typically given at the time to neutralize the shock of rape; it is equally indicated to neutralize the memory of it and its residual effects on the person's psyche. Next, to address the resultant emotions:

Humiliation, shame, being put down forcibly, cannot bear it	CHAM, COLOC, IGN, LYC, NAT-M, PH-AC, **STAPH**	*Acon, Anac, Arg-n, Ars, Aur, Nux-v, Puls, Sep, Sil, Stram, Sulph*	Bell, Calc, Lach, Merc, Plat

Looking at it two ways shows a good follow-up remedy in Staphysagria (STAPH):

Anger with indignation	COLOC, **STAPH**	*Aur, Nux-v*	Ars, Lyc, Merc, Nat-m, Plat
Anger with silent grief	IGN, LYC, **STAPH**	*Acon, Chin, Coloc, Nat-m, Ph-ac*	Bell, Carc, Cham, Nux-v, Plat, Puls

One dose of Stramonium 200C is given, after the patient has brought the event to conscious recall. Allowed to act for a week or so, a follow-up remedy in the form of Staphysagria 1M is given. After first neutralizing the terror of the event with the Stramonium, the psyche is primed to address the shame and anger with the Staphysagria.

At her next visit, she is feeling very sentimental about her father. She says she cries often about that but does not feel actually depressed. The anger she had been feeling is gone.

To act on the unresolved grief for her father:

Death of parents or friends	CAUST, IGN		Ars, Calc, Nit-ac, Nux-v, Plat, Staph

When checked to see if she has any of the confirming symptoms of Ignatia, it is found that she does have the characteristic "lump in the throat" sensation. A test for biocompatibility shows positive. She is given one dose of Ignatia 1M. At follow-up three weeks later, she returns complaining of increasing allergy and sinus problems. She wants to use allopathic medications to "dry up" her nasal passages and alleviate the symptoms, but the physician dissuades her and gives her Vitamin C with bioflavonoids to ease her discomfort.

Analysis

After moving backwards in time through the rape, anger, and grief issues, the condition that was present before that (sinus problems) re-surfaces. This phenomenon is all the more predictable, as the condition was allopathically suppressed for years with symptomatic medications. Old, suppressed conditions re-appear as the healing process works its way back through the health history. Within a short time, the symptoms abate. She also comments that her menstrual cycle has been regular for the last two months. The patient is having no overt problems and does not return for two months.

When she returns, her sex drive has returned, and she is emotionally stable. The reason for the visit is a persistent cough. It is dry and hoarse, and she presents an anxious appearance. It won't go away, she says, and she coughs so hard she fears that she'll not be able to get her breath. She emphasizes her anxiety over this, and consultation of the repertory reveals the following:

Fear of suffocation	ACON	*Ars, Phos, Staph, Stram, Sulph*	Lach, Merc, Nux-v

Knowing her early history of pneumonia, a test for appropriate medicines and causative agents (on lung test points) is warranted, and a positive reaction to Pneumococcus is found. Note that Staphysagria is indicated for fear of suffocation, but this time it does not match the case. While the fear is part of the symptom picture of Staphysagria, the physical characteristics of her cough do not fit the textbook description of the remedy—but it *does* fit Aconitum, the single medicine listed in Grade 3. She has the hard, croupy cough with anxiety, rapid pulse, and thirst that indicates this remedy. Her EDS, muscle test, or other test confirms this, and so Pneumococcinum D30 is given in a single dose, and Aconitum 200C is given every four hours for four doses. She calls the next day to say that there is a marked reduction of her cough. Note: Had this not been an old symptom returning, the prescription would likely have been a lower potency of Aconitum, since the illness would have been more on the superficial, physical level. The higher potencies are able to eradicate long-standing illness patterns.

Analysis

The healing has worked its way back in time to the age of four, and the latent pneumonia picture reveals itself. To neutralize the original causative agent, the nosode for the disease (in this case, Pneumococcinum) is given. Also, an appropriate medicine is given to act on the presenting symptoms (acting also as a drainage vehicle for the body to excrete the toxins associated with the old pneumonia), and it is given in a high enough potency to act quickly on the emotional component of the cough.

Following this, the patient has no ongoing problems but a month or two later, she is given two doses of Carcinosinum 200C to address the family history of cancer, but also the lingering fear of cancer in the patient. As we see, both can be dealt with this medicine:

Fear of cancer	ARS, CALC, **CARC**, PHOS, PLAT,	*Nit-ac*	Bar-c, Ign, Med, Nat-m, Sep

Summary

The depression in the start of this case was simply the end result of a string of trauma and suppressed symptoms. Sequentially addressing each of them *in reverse order of appearance in time,* as Hering's Law dictates, has the best chance of truly curing the patient. Treating the depression symptoms with an allopathic anti-depressant would have predictably resulted in an increase in allergy symptoms. Treating those with decongestants and anti-histamines would have resulted in further derangement of the menstrual cycle and uterine problems. Over time, the buried anger combined with the familial disposition toward cancer could have resulted in tumors in either the uterus or the site of the old unresolved pneumonia in the lung. Homeopathic medical literature describes thousands of cases with similar aspects.

Just as psychotherapy often seeks to uncover old trauma and bring them to conscious recall in order to neutralize their effects on the person's current mind state, treatment with high-potency homeopathic medicines can do likewise. With the Neo-eclectic approach, there is the added advantage of being able to use diagnostic signs from the body to indicate the appropriate medicine for each stage of healing.

Treatment should always be aimed at *the most strongly felt emotion* for each level. On one it may be anger; another, sadness; fear in a third stage, etc. As physical detoxification symptoms arise, treat with appropriate methods to insure adequate drainage of the toxins being expelled.

To start getting a feel for the psychological state of the patient, it is suggested that the clinician ask these questions or ones like them (reiterated from Chapter Three):

- What emotions do you have the most trouble dealing with?
- Do you have any frustrated ambitions?
- What characteristics prevent you from being exactly the person you would like to be?
- What creates the greatest stress in your life?
- How would a friend describe you?
- How would someone who does not like you describe you?
- What thing makes you the angriest?
- What thing makes you the saddest?
- What is the scariest thing in the world to you?

- What do you find yourself thinking about more than anything else?
- What do you worry about the most?
- What makes you the happiest?
- What kind of things do you do rapidly?
- What kind of things do you do slowly?

The things the examiner must privately decide are:
- What is to be cured in this patient?
- What is the basic dilemma of this person's life?
- Is there a strong sense of purpose, or apathy?
- Is there strength in the individual ego, or is the patient weak and unassertive?
- Is the patient able to express emotions with clarity, or are they vague or bottled up?

ANALYZING THE PSYCHOLOGICAL DISTURBANCE

The **psychoemotional repertory** in the next chapter is divided into categories dealing with
1. Personality traits
2. Persistent or chronic abnormal states
3. Causative factors of various types of trauma
4. Fears
5. Delusions, illusions, and ideations
6. Sleep disturbances

- When the problem lies in some pronounced trait of the patient's normal personality, use Section 1. Predominant mood or eccentricity can be a useful guide to the effective medicine to correct a psychological disturbance.

- When the problem lies in some *deviation* from the normal personality, thought processes, or emotional expression, use Section 2.

- Use Section 3 to define the traumatic factors leading to the predominant problem.

- When there are irrational fears or phobias, use Section 4, and add the one most strongly felt fear to the analysis.

- Delusions, illusions of the various senses, or persistent thoughts or ideations are very important symptoms to include in the analysis and are listed in Section 5.

- If sleep disturbance is a major symptom of the patient's psychoemotional problem, include it as defined in Section 6.

The repertory is further divided into three gradings, according to the traditional homeopathic methodology. Next to each symptom are three categories of indicated medicines. Grade 3 is the far left and those medicines are listed in all capitals. They are those with the strongest correlation with the symptom. Grade 2 medicines are slightly less strongly indicated by the specific symptom but may well include the appropriate choice for the patient. These are indicated by italics. Grade 1 contains the medicines that are least strongly indicated for that symptom but are still able to exert an influence over it. These are listed in lower case.

The medicine abbreviations are as listed in *Kent's Repertory of Homeopathic Materia Medica*. Although there are variations, we have conformed to the Kent abbreviations as they are the most well known.

This psychoemotional repertory aims at rendering the symptoms in a more contemporary form so as to eliminate some of the frustrations in trying to analyze cases using Kent's and others' texts of a hundred years ago. For example, only someone familiar with Kent's repertory would know to analyze a case of obsessive-compulsive disorder by looking up "Trifles, conscientious about". A modern-day user must internally translate his current language and the language of his patient into the 19th Century dialect of the older books. This will not be the case here. We have strived to make this work efficient for the contemporary user who does not have the advantage of years of experience with older texts.

By assembling the most pertinent symptoms for the stage of the patient's treatment at hand, then correlating the lists of indicated medicines for those symptoms, one will see certain medicines appear multiple times. This narrows down the list of possible effective medicines. Checking the symptom picture in a homeopathic materia medica to see if it matches the symptom picture of the patient is the next usual step. Then the medicines can be tested (by EDS, muscle testing, heart rate variability, etc.) to see which have the most profoundly therapeutic action or biocompatibility. This often disqualifies a medicine that otherwise looks good theoretically.

When medicines are given with this kind of precision, they have a dramatically corrective action.

The recommended administration is to start with a 200C potency of the indicated medicine, given only once or twice a day for one to two days. Wait two to three weeks to see if there has been a change. If there is a positive change, even if subtle, give the same medicine in a single dose of a higher potency (usually 1M). Advance to a 10M or 50M potency only if symptoms do not clear completely, or if they relapse and are not better from repeating the lower potency.

ASSESSING RESPONSE

Positive response to treatment is indicated by combinations of the following observations:

1. Being able to deal with stress more effectively.
2. Increased energy; able to do more.
3. Feeling better within oneself, more confident. Even if symptoms are still present, a feeling of being more like one's old self is a certain sign of improvement.
4. Increased creativity.
5. Making life changes. A change in relationship, job, or simply a change in attitudes regarding them, is a sign of improvement. The patient should realize that being rid of bad relationships and/or making new ones is often an important part of getting better—something that will not necessarily happen if taking allopathic drugs for the depression, anxiety, or whatever the problem is.
6. The classic homeopathic aggravation: The medicine provokes a healing reaction that accurately targets the core problem, causing a temporary worsening of symptoms. A return of the problem in its original form may occur; however, it will not be in the original intensity. Moreover, if counseling is done at this time to help the patient process the changes, improvement will progress smoothly. Professional counseling is not necessarily the only option. A close friend or relative may be all that is needed. **The homeopathic medicine must not be repeated** when this occurs. If allowed to act without repetition, the case will improve the fastest and smoothest. Only repeat the medicine if the patient regresses after experiencing clear improvement.
7. No change at all? The medicine given was likely inaccurate. Re-analyze the case and give the next most likely medicine.

Bibliography

- Dewey, W., *Practical Homeopathic Therapeutics,* Boericke & Tafel, 1901
- Dorchester, F., *Psycho-Physio-Kinesiology*, The Christopher Publishing House, 1928
- Kent, J., *Repertory of the Homeopathic Materia Medica,* 5th Edition, Ehrhart & Karl, 1945
- Lindlahr, H., *The New Psychology*, Lindlahr Pub. Co., 1910
- Lindlahr, H., *Mental, Emotional and Psychic Disorders*, Lindlahr Pub. Co., 1918
- Murphy, R., *Homeopathic Medical Repertory*; Hahnemann Academy of North America, 1993
- Powell, M., *An Outline of Naturopathic Psychotherapy,* 1st Edition, British College of Naturopathy and Osteopathy, 1967
- Reckeweg, H-H., *Homotoxicology,* 3rd Edition, Menaco Pub., Aurelia-Verlag 1991
- Smith, T., *Homeopathic Medicine for Mental Health,* Healing Arts Press 1989

15

PSYCHOEMOTIONAL REPERTORY

SECTION 1.
PERSONALITY TRAITS, ATTITUDES AND BEHAVIOR,
pp. 243 - 296

SECTION 1.
PERSONALITY TRAITS, ATTITUDES AND BEHAVIOR

These characteristics are ones that define a person and stand out so prominently that a number of people would agree that they describe the person. Consider the ones that are most appropriate for arriving at a prescription.

EXTERNALIZED BEHAVIOR TOWARD OTHERS			
Abruptness of manner	PULS	*Plat, Tarent*	Calc, Cham, Lyc, Med, Nat-m, Nit-ac, Nux-v, Sil, Sulph
Abusive, insulting	CHAM, LYC,	*Acon, Anac, Arn, Aur, Bell, Hyos, Ign, Nux-v, Sep, Stram*	Tub
Abusive, children insulting parents	PLAT		Hyos, Lach, Lyc, Nat-m
Addictive personality (drugs, alcohol, etc.)		*Alco, Carc, Lach, Med, Nux-v, Op, Thuj*	Ars, Calc, Lach, Ph-ac, Sulph
Anarchism	MERC		Arg-n, Caust
Apathetic (See *Indifference*)			
Apathy (See *Indifference*)			

EXTERNALIZED BEHAVIOR TOWARD OTHERS			
Arrogant	LYC, PLAT, SULPH, VERAT	*Caust, Ip, Lach, Pall, Puls, Stram*	Acon, Agar, Alum, Anac, Arn, Asar, Aur, Bell, Calc, Cann-i, Cann-s, Chin, Cic, Con, Cupr, Dulc, Ferr, Ign, Kali-i, Lil-t, Merc, Nat-m, Nux-v, Phos, Sabad, Sec, Thuj
Bitterness, exasperation			Ars, Ign, Nit-ac, Puls, Sulph
Blames others	ACON, CHIN	*Calc, Carc, Hyos, Ign, Lach, Lyc, Med, Merc, Nat-m, Nux-v, Staph*	Cham, Caust, Sep, Sulph
Blames self	ARS, AUR, HYOS, IGN, NAT-M, PULS, SULPH	*Lach, Lyc, Med, Merc, Sil, Stram, Thuj*	Calc-p, Cob, Cycl, Hell, Hura, Lyc, Merc, Nat-a, Ph-ac
Bossy (See *Dictatorial*)			
Break things, desire to		*Nux-v, Stram, Tub*	Bell, Hyos, Staph, Sulph, Tarent

EXTERNALIZED BEHAVIOR TOWARD OTHERS			
Busy, must be	AUR, TARENT	*Bar-c, Croc, Hyos, Ign, Lach, Op, Sep*	Agar, Arn, Ars, Bell, Bry, Calc, Calc-p, Caps, Chin, Clem, Dig, Indg, Ip, Kreos, Led, Mag-c, Mez, Mosch, Mur-ac, Nat-c, Nat-s, Phos, Pip-m, Plb, Rhus-t, Stann, Sul-ac, Verat
Contemptuousness	PLAT	*Arg-n, Ars, Chin, Lyc, Nux-v*	
Contradiction, disposition to	ANAC, CAUST, LACH	*Ars, Aur, Lyc, Merc*	
Contrary, takes the opposite action as desired	ALUM, ANAC, ARG-N, CHAM, HEP, LACH, MERC, TARENT	*Ant-t, Arn, Ars, Bar-c, Caust, Cocc, Kali-c, Nit-ac, Nux-v, Puls, Sulph, Thuj*	Abrot, Acon, Ambr, Anan, Ant-c, Arum-t, Aur, Bell, Bry, Calc, Calc-s, Calc-sil, Camph, Canth, Caps, Carb-an, Chin, Cina, Con, Croc, Guaij, Ign, Ip, Kali-P, Kreos, Lact, Laur, Led, Lyc, Mag-C, Mag-M, Petr, Phos, Plb, Ruta, Samb, Sars, Sep, Sil, Spong

EXTERNALIZED BEHAVIOR TOWARD OTHERS			
Criticalness	ARS, SULPH	*Bar-c, Caust, Lach, Lyc, Nux-v, Phos, Plat, Sep*	
Cruelty, inhumanity	ANAC	*Ars, Hyos, Lach, Nit-ac, Staph, Stram*	Bell, Carc, Chin, Med, Nux-v, Tarent
Cruelty, to animals	ARS		Bell, Calc, Med
Destructiveness	STRAM	*Hyos, Lach, Sulph, Tarent*	Bell, Calc, Carc, Nux-v, Phos, Plat, Staph, Tub
Dictatorial	NAT-M, SEP	*Camph, Lyc, Merc*	Aur, Caust, Cham, Con, Ferr, Lach
Egotistical	LYC, PLAT, SULPH, VERAT	*Caust, Ip, Lach, Pall, Puls, Stram*	Acon, Agar, Alum, Anac, Arn, Asar, Aur, Bell, Calc, Cann-i, Cann-s, Chin, Cic, Con, Cupr, Dulc, Ferr, Ign, Kali-i, Lil-t, Merc, Nat-m, Nux-v, Phos, Sabad, Sec, Thuj
Enemy, considering everybody		Merc	

EXTERNALIZED BEHAVIOR TOWARD OTHERS			
Estrangement from family		*Nat-m, Nit-ac*	Anac, Arn, Ars, Nux-v, Phos, Plat, Sep, Staph
Flatterer	LYC		Arn, Carb-v, Nux-v, Petr, Plat, Puls, Sil, Staph, Sulph
Flattery, desires		*Pall*	Carb-v, Lyc, Puls
Hardheartedness, lack of feeling, cold hard look	ANAC	*Lach, Plat, Sulph*	Ars, Hyos
Hatred	ANAC, NAT-M, NUX-V	*Aur, Calc, Cham, Lach, Nit-ac, Ph-ac*	
Hatred, of men			Bar-c, Ign, Lyc, Phos
Hatred, of women			Puls
Haughty	LYC, PLAT, SULPH, VERAT	*Caust, Ip, Lach, Pall, Puls, Stram*	Acon, Agar, Alum, Anac, Arn, Asar, Aur, Bell, Calc, Cann-i, Cann-s, Chin, Cic, Con, Cupr, Dulc, Ferr, Ign, Kali-i, Lil-t, Merc, Nat-m, Nux-v, Phos, Sabad, Sec, Thuj

EXTERNALIZED BEHAVIOR TOWARD OTHERS			
Impetuous	CHAM, HEP, IGN, NIT-AC, NUX-V, SEP	*Anac, Bry, Carb-v, Cham, Kali-c, Med, Nat-m, Sul, Staph, Sulph, Zinc*	Acon, Caust, Croc, Ferr-p, Kali-i, Kali-p, Laur, Led, Nat-c, Olnd, Phos, Rheum, Stront
Indecisive	BAR-C, HELL, IGN, LACH, ONOS, OP	*Agar, Alum, Arg-n, Ars, Bar-m, Calc, Carb-s, Cocc, Con, Cur, Graph, Ip, Lyc, Merc, Mez, Naja, Nat-m, Nux-m, Nux-v, Petr, Phos, Psor, Puls, Sep, Sil, Sulph*	Am-c, Apis, Arn, Ars-I, Asaf, Aur, Bry, Cact, Calc-fl, Calc-p, Calc-s, Camph, Cann-i, Cann-s, Canth, Caust, Cham, Chin, Cina, Clem, Coca, Coff, Coll, Cupr, Dig, Dros, Dulc, Ferr, Hyos, Kali-ar, Kali-br, Kali-c, Kali-p, Kali-s, Lac-d, Laur, Led, Mag-m, Mang, Nat-c, Nit-ac, Pall, Pic-ac, Plat, Plb, Rheum, Rhus-r, Ruta, Sanic, Seneg, Spig, Tab, Ta-rax, Tarent, Thuj, Zinc

EXTERNALIZED BEHAVIOR TOWARD OTHERS			
Industrious	AUR, TARENT	*Bar-c, Croc, Hyos, Ign, Lach, Op, Sep*	Agar, Arn, Ars, Bell, Bry, Calc, Calc-p, Caps, Chin, Clem, Dig, Indg, Ip, Kreos, Led, Mag-c, Mez, Mosch, Mur-ac, Nat-c, Nat-s, Phos, Pip-m, Plb, Rhus-t, Stann, Sul-ac, Verat
Insulting	CHAM, LYC	*Acon, Anac, Arn, Aur, Bell, Hyos, Ign, Nux-v, Sep, Stram*	Tub
Intellectual	ACON, BAR-C, BELL, HELL, HYOS, LACH, LYC, OP, PH-AC, PHOS, SEP, STRAM, SULPH, VERAT	*Plat, Puls, Rhus-t,*	Anac, Aur, Bapt, Cann-i, Cocc, Ign, Laur, Merc, Nat-c, Nat-m, Nux-v, Sil
Kill, desire to	HYOS, PLAT, STRAM	*Ars, Hep, Iod, Lyc, Med, Merc, Nux-v, Phos, Sil, Staph*	

EXTERNALIZED BEHAVIOR TOWARD OTHERS			
Kleptomania	BELL, OP	*Absin, Art-v, Cur, Nux-v Puls, Sulph, Thuj*	Ars, Bry, Caust, Kali-c, Lyc, Sep, Sil, Staph, Stram, Sulph, Tarent
Laughing tendency	BOR, CANN-I, HYOS, IGN, PHOS, STRAM	*Arg-m, Ars, Aur, Bell, Bufo, Calc, Cann-s, Coff, Cupr, Ferr, Lach, Lyc, Nat-m, Nux-m, Plat, Sep, Stann, Tarent*	Acon, Agar, Alco, Aloe, Alum, Ambr, Am-c, Anac, Anan, Apis, Arn, Arund, Asaf, Aur-a, Bar-c, Caps, Carb-v, Caust, Cic, Con, Cor-r, Crot-h, Cypr, Ferr-ar, Graph, Hell, Hura, Hydrog, Kali-bi, Kali-p, Kreos, Lepi, Lil-t, Merl, Nat-a, Nat-c, Nux-v, Op, Plb, Puls, Ran-h, Rob, Sabad, Sant, Sarr, Scorp, Sec, Sil, Spong, Stry, Sulph, Sumb, Tab, Tarax, Valer, Verat, Verb, Zinc, Zinc-s
Lies, tries to appear truthful	MORPH, OP	*Alco, Syph, Verat*	
Loquacity (See **Talkativeness**)			

EXTERNALIZED BEHAVIOR TOWARD OTHERS			
Lying	MORPH, OP	*Alco, Lyc, Syph, Thuj, Verat*	Arg-n, Calc, Carb-v, Caust, Coca, Con, Merc, Nat-m, Nux-v, Puls, Scorp, Sil, Staph, Sulph
Mocking		*Lach*	Acon, Ars, Chin, Hyos, Ign, Nux-v, Plat, Tarent
Ordering (See **Dictatorial**)			
Personal appearance, careful about	ARS, CARC,	*Anac, Kali-c, Med, Nat-m, Nux-v*	Arg-n, Phos, Plat, Puls, Sep, Sil, Sulph, Thuj
Pompous, feels overly important	PLAT	*Lyc, Nux-v, Sulph, Verat*	Bell, Calc, Cupr, Phos
Quarrelsomeness	AUR, HYOS, IGN, NUX-V, SULPH, TARENT	*Anac, Arn, Ars, Bell, Caust, Cham, Kali-c, Lach, Lyc, Merc, Nat-m, Nit-ac, Ph-ac, Phos, Plat, Sep, Sil, Staph, Stram, Thuj*	

EXTERNALIZED BEHAVIOR TOWARD OTHERS			
Rage, fury	BELL, HYOS, LYC, STRAM	*Acon, Anac, Arn, Ars, Lach, Merc, Nat-m, Nit-ac, Phos, Puls, Sulph*	
Rebelling, revolu-tionary	MERC	*Caust*	Kali-c, Sep
Resentment, vengefulness and maliciousness	ANAC, ARS, NAT-M, NUX-V, STRAM, TUB	*Acon, Aur, Bell, Calc, Cham, Hyos, Lach, Lyc, Nit-ac, Ph-ac, Staph*	
Revenge (See ***Vengeful***)			
Ridicule, tendency to			Acon, Hyos, Lach, Nux-v
Rudeness	HYOS, PLAT	*Anac, Cham, Chin, Nux-v, Phos, Stram*	Lyc
Swearing	NIT-AC	*Ars, Bell, Hyos, Lyc, Nat-m, Nux-v, Tub*	
Sympathy, lack of			Anac, Ars, Cham, Chin, Nat-m, Nit-ac, Plat, Puts, Sep
Talk, indisposed to	AUR, CARB-AN, COCC,	*Acon, Agar, Ant-c, Arg-m, Arg-n, Arn,*	Abrot, Alco, Alum, Am-c, Am-m, Anac, Ant-t, Apoc,

EXTERNALIZED BEHAVIOR TOWARD OTHERS			
Talk, indisposed to, cont'd.	GLON, PH-AC, PHOS, PLAT, PULS, SULPH, VERAT, ZINC	*Ars, Arund, Bar-c, Bell, Calc, Caps, Carb-v, Caust, Chin, Cimic, Coloc, Euph, Ferr, Gels, Hell, Hep, Hipp, Hyos, Ign, Lyc, Lycps, Mag-m, Mang, Merc, Mur-ac, Nat-m, Nat-s, Nit-ac, Nux-v, Pic-ac, Plb, Rhus-t, Scorp, Stann, Staph, Stram, Tarent, Thuj*	Aster, Bapt, Bar-m, Berb, Bor, Brom, Bry, Bufo, Cact, Calc-p, Calc-s, Cann-i, Cann-s, Canth, Carb-ac, Carb-s, Carc, Cham, Chel, Cic, Cina, Clem, Colch, Crot-t, Cupr, Cycl, Dig, Dros, Euphr, Fl-ac, Graph, Grat, Guai, Ham, Helon, Hydr, Iod, Ip, Piloc, Kali-ar, Kali-bi, Kali-c, Kali-m, Kali-p, Kali-s, Kali-Sil, Kreos, Lac-ac, Lach, Led, Lil-t, Mag-c, Mag-s, Manc, Many, Mez, Mosch, Murx, Naja, Nat-a, Nat-c, Nat-p, Nux-m, Onos, Op, Ox-ac, Petr, Phys, Sabin, Sec, Sep, Sil, Spig, Spong, Stront-c, Sul-ac, Tarax, Tub, Ust, Zinc-p

EXTERNALIZED BEHAVIOR TOWARD OTHERS			
Talkativeness (loquacity), or tendency to write long letters	HYOS, LACH, STRAM	*Agar, Arg-m, Aur, Bell, Cann-i, Cimic, Cocc, Cupr, Gels, Iod, Kali-I, Mosch, Mur-ac, Nat-c, Op, Par, Phos, Plb, Podo, Pyrog, Verat*	Abrot, Acon, Agn, Alco, Ant-t, Apis, Ars-i, Bar-c, Bar-I, Calc, Caust, Chel, Coff, Crot-h, Dulc, Glon, Hep, Lil-t, Lyc, Nat-a, Nat-m, Nux-m, Nux-v, Psor, Rhus-t, Staph, Sulph, Tarax, Tarent, Thea, Ther, Thuj, Valer, Verat-v, Zinc
Vengeful	ANAC, ARS, NAT-M, NUX-V, STRAM, TUB	*Acon, Aur, Bell, Calc, Cham, Hyos, Lach, Lyc, Nit-ac, Ph-ac, Staph*	

INTERNALIZED BEHAVIOR			
Alone, desire to be	ANAC, BAR-C, CARB-AN, CHAM, CIC, GELS, IGN, NAT-M, NUX-V	*Ambr, Aur, Bel, Bry, Cact, Calc-p, Carb-v, Chin, Coloc, Cupr, Cycl, Ferr, Hell, Hep, Hyos, Iod, Lach, Led, Lyc, Nat-c, Plat, Puls,*	Acon, Alum, Anan, Ant-c, Ant-t, Bufo, Calc, Cann-i, Carb-s, Cedr, Cimic, Clem, Coca, Con, Cop, Dig, Dios, Elaps, Fl-ac, Graph, Grat, Ham, Helon, Hydr, Kali-bi, Kali-br, Kali-c,

INTERNALIZED BEHAVIOR			
Alone, desire to be, cont'd.		*Rhus-t, Sel, Sep, Stann, Sulph, Thuj*	Kali-p, Kali-s, Mag-m, Mang, Nat-p, Nicc, Petr, Phos, Pic-ac, Psor, Sul-ac, Tarent, Til, Verat
Anxious	ACON, ARG-N, ARS, ARS-I, AUR, BELL, BISM, BRY, CACT, CALC, CALC-P, CALC-S, CAMPH, CANN-I, CARB-S, CARB-V, CAUST, CON, DIG, IOD, KALI-AR, KALI-C, KALI-P, KALI-S, LYC, MEZ, NAT-A, NAT-C,	*Abrot, Acet-ac, Aeth, All-c, Alum, Ambr, Am-c, Anac, Ant-c, Ant-t, Arg-m, Arn, Asar, Bar-c, Bar-m, Bor, Bov, Canth, Carb-an, Carbo-o, Cham, Chel, Chin-a, Chin-s, Cic, Cimic, Cocc, Coc-c, Coch, Coff, Coloc, Crot-h, Cupr, Dros, Euph, Ferr, Fl-ac, Gels, Graph, Hell, Hep, Hyos, Ign, Kali-i, Kali-n, Lach, Laur, Led,*	

INTERNALIZED BEHAVIOR		
Anxious, cont'd.	NIT-AC, PHOS, PULS, RHUS-T, SEC, SULPH, VERAT	*Lil-t, Mag-c, Mag-m, Mag-s, Merc, Mur-ac, Nat-p, Nat-m, Nux-v, Op, Petr, Plb, Ruta, Sa-bad, Sabin, Samb, Seneg, Staph, Stram, Tab, Tar-ent, Thuj, Zinc*	
Bad luck, feeling of	LYC	*Chin, Sep, Staph*	Carc, Kali-c, Sulph
Biting nails, pulling hair out	ACON, ARS	*Bar-c, Hyos, Lyc, Med, Nat-m, Stram, Sulph, Tar-ent*	Calc, Carc, Caust, Nit-ac, Phos, Puls, Sil, Staph
Biting objects	BELL, HYOS, STRAM	*Canth*	Sil
Blames self		*Acon, Ars, Aur, Dig, Hyos, Ign, Nat-m, Op, Puls, Sarr, Thuj*	Calc-p, Cob, Cycl, Hell, Hura, Lyc, Merc, Nat-a, Ph-ac

INTERNALIZED BEHAVIOR

Company, aversion to	ANAC, BAR-C, CARB-AN, CHAM, CIC, GELS, IGN, NAT-M, NUX-V	*Ambr, Aur, Bel, Bry, Cact, Calc-p, Carb-v, Chin, Coloc, Cupr, Cycl, Ferr, Hell, Hep, Hyos, Iod, Lach, Led, Lyc, Nat-c, Plat, Puls, Rhus-t, Sel, Sep, Stann, Sulph, Thuj*	Acon, Alum, Anan, Ant-c, Ant-t, Bufo, Calc, Cann-i, Carb-s, Cedr, Cimic, Clem, Coca, Con, Cop, Dig, Dios, Elaps, Fl-ac, Graph, Grat, Ham, Helon, Hydr, Kali-bi, Kali-br, Kali-c, Kali-p, Kali-s, Mag-m, Mang, Nat-p, Nicc, Petr, Phos, Pic-ac, Psor, Sul-ac, Tarent, Til, Verat
Crying, tearful mood	APIS, CALC, CALC-S, CARB-S, CAUST, GRAPH, IGN, KALI-BR, LAC-C, LYC, NAT-M, PLAT, PULS, RHUS-T, SEP, SULPH, VERAT	*Acon, Alum, Am-m, Arg-n, Aur, Bell, Calc-p, Carb-v, Cham, Chel, Chin-s, Cimic, Coff, Con, Cupr, Dig, Ferr, Hell, Hep, Iod, Kali-bi, Kali-c, Kali-p, Lil-t, Mag-m, Mang, Med, Nat-p, Nat-s,*	Am-c, Ars, Bar-c, Bor, Cann-i, Colch, Dulc, Gels, Glon, Hura, Lith, Mosch, Nat-c, Psor, Sabin, Sec, Spig, Stram, Thuj, Zinc

INTERNALIZED BEHAVIOR			
Crying, tearful mood, cont'd.		*Nit-ac, Nux-m, Ph-ac, Phos, Spong, Staph, Sul-ac, Tarent, Viol-o*	
Dependent, co-dependency, etc.	AUR, HYOS, PH-AC, PHOS, STAPH	*Bell, Caust, Gels, Ign, Kali-c, Lach, Lil-t, Lyc, Nux-v, Puls, Sep, Tarent*	Arg-n, Ars, Brom, Bry, Calc, Calc-p, Camph, Clem, Con, Crot-h, Dros, Hep, Kali-p, Pall, Stram, Verat, Zinc
Depression, de-spair, hopeless	ARS, AUR, CALC, IGN, LYC, SEP, SULPH	*Anac, Arg-n, Cann-i, Caust, Chin, Lach, Med, Merc, Nat-m, Nit-ac, Nux-v, Puls, Staph, Stram*	
Depression, suici-dal disposition from	AUR, SEP	*Anac, Ars, Bell, Calc, Carc, Chin, Hyos, Ign, Lach, Med, Merc, Nat-m, Nux-v, Puls,Stram*	
Despair, caused by pain	AUR, IGN, NAT-M,	*Ars, Cham, Chin*	Acon, Calc, Nux-v, Stram

INTERNALIZED BEHAVIOR			
Despair, of recovering	ARS, AUR, CALC, MERC	*Ign, Med, Nit-ac, Sep*	Acon
Despair, religious	ARS, AUR, LACH, PULS	*Arg-n, Calc, Lyc, Med, Stram, Sulph, Thuj*	
Discontented	ANAC, CALC-P, MERC, NAT-M, SULPH	*Am-m, Ars, Aur, Bism, Bor, Bry, Cham, Chel, Chin, Cina, Colch, Cupr, Hep, Kali-c, Lyc, Nit-ac, Nux-v, Pall, Plat, Puls, Rhus-t, Sep, Sil, Stann, Staph, Thuj*	Acon, Aeth, Agar, Agn, Alet, Aloe, Alum, Am-c, Apis, Arn, Ars-i, Asar, Aur-m, Bar-c, Bell, Berb, Bov, Brom, Calc, Calc-s, Cann-s, Canth, Caps, Carbo-an, Carb-s, Caust, Chin-a, Cic, Clem, Cocc, Coff, Coloc, Con, Crot-t, Dulc, Ferr, Ferr-p, Fl-ac, Graph, Grat, Ham, Hell, Hura, Ign, Indg, Iod, Ip, Kreos, Lach, Laur, Led, Lil-t, Mag-c, Mag-m, Mag-s, Manc, Mang, Mez, Mur-ac, Nat-a, Nat-c, Nat-p, Nit-ac, Op,

INTERNALIZED BEHAVIOR			
Discontented, cont'd.			Orig, Par, Petr, Ph-ac, Phos, Plb, Ran-b, Rhod, Rob, Ruta, Samb, Sars, Spong, Tarent, Ther, Til, Viol-t
Dissatisfied (See **Discontented**)			
Disgust, feeling of	PULS, SULPH	*Kali-c, Merc, Stram*	
Doubt		*Lach*	Ars, Aur, Bar-c, Chin, Nux-v, Sep, Sil, Staph, Thuj
Escape, desire to	BELL, HYOS	*Agar, Ars, Bry, Cocc, Crot-h, Cupr, Dig, Glon, Nux-v, Op, Stram, Verat*	Acon, Aesc, Ars-m, Arum-t, Bapt, bar-c, Caust, Cham, Chin, Cic, Coloc, Hell, Ign, Lach, Lil-t, Meli, Merc, Merc-c, Phos, Plb, Puls, Ran-b, Rhus-t, Samb, Sol-n, Sulph, Sul-ac, Tub, Zinc

INTERNALIZED BEHAVIOR			
Escapist tenden-cies	BELL, HYOS	*Agar, Ars, Bry, Cocc, Crot-h, Cupr, Dig, Glon, Nux-v, Op, Stram, Verat*	Acon, Aesc, Ars-m, Arum-t, Bapt, bar-c, Caust, Cham, Chin, Cic, Coloc, Hell, Ign, Lach, Lil-t, Meli, Merc, Merc-c, Phos, Plb, Puls, Ran-b, Rhus-t, Samb, Sol-n, Sulph, Sul-ac, Tub, Zinc
Guilt	ARS, AUR, SULPH	*Bell, Caust, Hyos, Ign, Lach, Med, Merc, Nat-m, Nux-v, Ph-ac, Puls, Sil, Stram*	Thuj
Hanging, suicidal disposition	ARS	*Bell*	Aur
Hopeless	ARS, AUR, CALC, MERC	*Ign, Med, Nit-ac, Sep*	Acon
Killed, desire to be		*Ars*	Bell, Stram

INTERNALIZED BEHAVIOR			
Obsessive-compulsive disorder (See also **One-track mind**, **Repetitive Actions**, **Repetitive Thoughts**, **Washing**, etc.)	ARS, HYOS, IGN, MED, NUX-V, PULS, STAPH, THUJ	*Anac, Arg-n, Bar-c, Calc, Carc Lach, Lyc, Med, Nat-m, Nat-s, Plat, Sil, Stram, Sulph, Verat*	Aur, Iod, Syph, Psor, Sulph
Overwhelmed by needs or responsibilities	ARS, CALC, LYC	*Aur, Carc, Med, Phos-ac*	Anac, Calc-p, Fl-ac, Ign, Kali-c, Nat-m, Nat-s, Sulph
Perfectionist, compulsive about details	ARS, IGN, STAPH, THUJ	Bar-c, *Lach, Lyc, Med, Nux-v, Stram Sulph*	
Pleasure, loss of (no joy in formerly pleasant things)	HELL, NAT-M, OP, SULPH	*Arg-n, Ars, Cham, Chin-s, Puls, Sep*	Alum, Anac, Cocc, Croc, Ferr-p, Hep, Hura, Kali-ar, Kali-c, Mag-m, Mez, Mur-ac, Nat-a, Nat-c, Nit-ac, Petr, Sars, Spig, Stann, Staph, Stram, Tab, Ther
Sadness, suicidal disposition	ALUM, AUR, ARG-N, CHIN, NAT-S, VERAT	*Nat-m, Nux-v*	Calc, Caust, Chin, Ign, Med, Sep, Sulph

INTERNALIZED BEHAVIOR			
Self-blame	ARS, AUR, HYOS, IGN, NAT-M, PULS, SULPH	*Lach, Lyc, Med, Merc, Sil, Stram*	Thuj
Sensitive to others' criticisms	ARG-N, BELL, BOR, CHIN, COFF, GELS, IGN, LYC, LYSS, NAT-M, NIT-AC, NUX-V, PHOS, PLB, PULS, RAN-B, SIL, SULPH, THER, VALER	*Acon, Aesc, Arn, Ars, Aur, Bar-c, Calc, Canth, Carb-s, Carb-v, Caust, Cham, Cocc, Crot-h, Ferr, Fl-ac, Hyos, Iod, Kali-ar, Kali-c, Kali-i, Kali-p, Lac-c, Lach, Mag-m, Med, Merc, Mosch, Nat-a, Nat-c, Nat-p, Nat-s, Plat, Sabin, Seneg, Sep, Staph, Zinc*	Alum, Am-c, Anac, Apis, Asaf, Bry, Calc-p, Cann-s, Cab-an, Cic, Cina, Clem, Colch, Coloc, Con, Cupr, Daph, Dig, Dros, Hep, Kreos, Mez, Ph-ac, Psor, Sabad, Sanic, Sars, Stann, Tab, Thuj, Verat
Self-pity	CALC	*Staph*	Nit-ac
Seriousness, earnestness	ARS	*Aur, Chin, Merc, Staph*	

INTERNALIZED BEHAVIOR			
Smiling, never	ARS, AUR		
Solitude, desire for	ANAC, BAR-C, CARB-AN, CHAM, CIC, GELS, IGN, NAT-M, NUX-V	*Ambr, Aur, Bel, Bry, Cact, Calc-p, Carb-v, Chin, Coloc, Cupr, Cycl, Ferr, Hell, Hep, Hyos, Iod, Lach, Led, Lyc, Nat-c, Plat, Puls, Rhus-t, Sel, Sep, Stann, Sulph, Thuj*	Acon, Alum, Anan, Ant-c, Ant-t, Bufo, Calc, Cann-i, Carb-s, Cedr, Cimic, Clem, Coca, Con, Cop, Dig, Dios, Elaps, Fl-ac, Graph, Grat, Ham, Helon, Hydr, Kali-bi, Kali-br, Kali-c, Kali-p, Kali-s, Mag-m, Mang, Nat-p, Nicc, Petr, Phos, Pic-ac, Psor, Sul-ac, Tarent, Til, Verat
Suicidal disposition	ALUM, AUR, ARG-N, CHIN, NAT-S, VERAT	*Nat-m, Nux-v*	Calc, Caust, Chin, Ign, Med, Sep, Sulph
Throwing oneself from a height, suicidal disposition	AUR, BELL	*Arg-n, Nux-v*	Anac, Ars, Hyos, Ign, Lach, Sil, Staph, Stram, Sulph, Thuj
Weariness with life	ARS, AUR, CHIN, NAT-M, PHOS	*Merc, Nit-ac, Nux-v, Ph-ac, Puls, Sil*	

SELF-ESTEEM, LOW			
Bashful (See **Shyness, strong**)			
Blames self		*Acon, Ars, Aur, Dig, Hyos, Ign, Nat-m, Op, Puls, Sarr, Thuj*	Calc-p, Cob, Cycl, Hell, Hura, Lyc, Merc, Nat-a, Ph-ac
Childish behavior	BAR-C	*Arg-n, Ign, Ph-ac, Stram*	
Confidence, lack of	ANAC, SIL	*Aur, Bry, Chin, Kali-c, Lyc, Med,* *Nat-m, Ph-ac, Puls*	Am-c, Alum, Bar-c, Calc, Chin, Nit-ac, Phos, Staph
Cowardice	LYC, STRAM	*Acon, Arg-n,* Bar-c, *Calc, Cham, Chin,* Nux-v, *Puls,* Sil	
Flattery of others	LYC		Arn, Nux-v, Plat, Puls, Sil, Staph
Helplessness, feeling of	LYC		Anac, Arg-n, Phos, Puls, Stram

SELF-ESTEEM, LOW			
Indecision, great	BAR-C, IGN, LACH	*Anac, Arg-n, Ars, Calc, Lyc, Merc, Nat-m, Nux-v, Phos, PULS, Sep, Sil, Sulph*	
Inferiority, feelings of	AUR, NAT-M, PULS	*Acon, Ars, Dig, Hyos, Ign, Op, Sarr, Thuj*	Calc-p, Cob, Cycl, Hell, Hura, Lyc, Merc, Nat-a, Ph-ac
Isolation, feeling of	STRAM	*Anac, Arg-n, Cann-i, Plat, Puls*	
Meekness, mildness	ARN, ARS, CARC, NAT-M, PULS, SIL	*Acon, Calc, Ign, Lyc, Nit-ac, Nux-v, Phos, Sep, Stram, Sulph, Thuj*	
Rejection, abandonment, desertion, feeling of	AUR, PULS, STRAM	*Arg-n Lach, Merc, Plat*	Anac, Bar-c, Calc, Cann-i, Carc, Chin, Kali-c, Sep
Reverence for others			Hyos, Nat-m, Nux-v, Plat, Puls, Sil, Sulph

Shyness	BAR-C, CALC, COCA, KALI-C, GELS, KALI-C, LYC, NAT-C, PETR, PHOS, PLB, PULS, SEP, SIL, SULPH	*Acon, Alum, Am-m, Ars, Aur, Caust, Chin, Coca, Con, Cupr, Graph, Ign, Kali-ar, Kali-s, Merc, Nat-m, Nit-ac, Nux-v, Rhus-t, Sil, Spong, Staph, Stram, Tub*	Ambr, Am-c, Anac, Arn, Ars-I, Bell, Bry, Canth, Carb-an, Cocc, Hyos, Kali-br, Lil-t, Mag-c, Mur-ac, Nat-p, Nit-ac, Plat, Ran-b, Sec, Spig, Sul-ac, Tab, Verat, Zinc
Shyness, strong; with poor eye contact, awkwardness, etc.	COCA, PULS	*Ambr, Bar-c, Calc, Carb-an, Chin, Cupr, Ign, Merc, Nat-c, Petr, Staph, Stram, Sulph, Zinc*	Anac, Arg-n, Aur, Bell, Carb-v, Con, Coff, Hyos, Iod, kali-bi, Kali-p, Manc, Mang, Mez, Nit-ac, Nux-v, Phos, Tab
Submissiveness	PULS	*Ars, Ign, Lyc, Nux-v*	Anac, Lyc, Sil, Staph, Sulph
Success, lack of, expectation of failure		*Arg-n*	Aur, Merc, Nux-v, Sil
Timidity (See **Shyness)**			

SELF-ESTEEM, INFLATED			
Boasting, bragging			Arn, Ars, Bell, Lach, Merc, Nat-m, Nux-v, Plat, Stram, Sulph
Hypocrisy		*Sil, Sulph*	BAR-C, Caust, Lyc, Merc, Nux-v, Phos, Puls, Sep
Bossiness	ARS	*Lyc*	
Eccentricity	LACH	*Bell, Cann-i, Tarent*	
Exaltation of politics		*Caust*	Bell, Lach, Nux-v
Egotism	PLAT	*Calic, Lach, Lyc, Sulph*	Anac, Arn, Aur, Med, Merc, Nux-v, Phos, Sil, Staph, Stram
Haughtiness	LYC, PLAT, SULPH	*Caust, Hyos, Lach, Puls, Sil, Staph, Stram*	
Power, love of		*Lyc*	Ars, Lach, Nux-v, Sulph
Pride	LACH, PLAT		Arn, Ars, Caust, Chin, Hyos, Lyc, Nux-v, Staph, Stram, Sulph
Presumptuousness		*Lyc*	Arn, Calc, Plat, Staph

SELF-ESTEEM, INFLATED			
Selfishness		*Calc, Puls, Sulph, Tarent*	Ars, Bell, Chin, Ign, Lach, Lyc, Med, Merc, Nat-m, Nux-v, Phos, Plat, Sep, Sil

CONTROL			
Carefulness, takes great care with everything	CHIN, STRAM, SULPH	*Ars, Aur, BAR-C. Ign, Lach, Lyc, Nux-v. Sep*	
Cautiousness		*Ign, Puls*	Ars, Bar-c, Calc, Caust, Hyos, Nux-v, Stram
Exacting, tendency to be too	ARS, HYOS, IGN, MED, NUX-V, PULS, STAPH, THUJ	*Anac, Arg-n, Bar-c, Calc, Carc Lach, Lyc, Med, Nat-m, Nat-s, Plat, Sil, Stram, Sulph, Verat*	Aur, Iod, Syph, Psor, Sulph
Conscientiousness about trifles (attention to small details but not necessarily obsessive-compulsive—See **Obsessive, Compulsive**)	ARS, IGN, STAPH, THUJ	*Bar-c, Lach, Lyc, Med, Nux-v, Stram Sulph*	

CONTROL			
Obsessive, compulsive about details	ARS, HYOS, IGN, MED, NUX-V, PULS, STAPH, THUJ	*Anac, Arg-n, Bar-c, Calc, Carc Lach, Lyc, Med, Nat-m, Nat-s, Plat, Sil, Stram, Sulph, Verat*	Aur, Iod, Syph, Psor, Sulph
Personal appearance, careful about	ARS, CARC	*Anac, Kali-c, Med, Nat-m, Nux-v*	Arg-n, Phos, Plat, Puls, Sep, Sil, Sulph, Thuj
Tidiness	ARS, CARC	*Anac, Kali-c, Med, Nat-m, Nux-v*	Arg-n, Phos, Plat, Puls, Sep, Sil, Sulph, Thuj

TRUST			
Betrayal	AUR, LYC, NAT-M	*Nux-v*	Ign, Lach, Merc, Ph-ac, Puls, Sep
Betrayal of ambition		*Nux-v*	Bell, Merc, Plat, Puls
Betrayal of confidence	ARG-N, AUR, NAT-C, PSOR, PULS	*Staph*	
Betrayal of friendship	ARG-N, AUR, NAT-C, PSOR, PULS	*Staph*	Ign, Nux-v, Ph-ac, Sil, Sulph

270

TRUST			
Suspicious (See *Trust, lack of*)			
Trust, lack of, suspicion	ACON, ANAC, ARS, BAR-C, CANN-I, CAUST, LACH, LYC, PULS, STRAM	*Arn, Aur, Med, Merc, Nit-ac, Nux-v, Phos, Sep, Staph*	Bell, Hyos

GUARDEDNESS			
Hypervigilance, Post-traumatic Stress Disorder (PTSD), etc.	ARN, ARS, STRAM	*Arg-n, Ign, Lach*	Arn, Aur, Bell, Cann-i, Chin, Hyos, Kali-c, Thuj
Reserve	NAT-M, PHOS	*Calc, Hyos, Ign, Plat, Puls, Staph*	
Secretiveness		*Bar-c, Lyc, Nat-m, Sep, Thuj*	Aur Caust, Ign, Nit-ac, Phos
Silent, taciturn, won't answer questions		*Ant-c, Bry, Cham, Col, Gels, Nat-m, Nux-v Ph-ac*	Ant-t, Arn, Bell, Cact, Carbo-an, Hell, Ign, Iod, Mur ac, Naja, Nat-s, Phos, Sars, Sil, Sul
Sulking, sullen, morose, unsociable		*Ant-c, Bry, Cham, Chin, Nux-v, Plat, Puls, Ver a*	Ant-t, Arn, Aur, Cimicic, Col, Con, Cupr, Ign, Lyc, Nat-m, Sil, Sul, Tub

OPENNESS			
Extroversion		*Lach, Phos*	Acon, Arg-n, Bar-c, Carc, Lyc, Med, Nux-v, Sulph, Tar-ent
Naivety			Bell, Stram
Naivety with great intelligence		*Chin, Hyos, Stram, Sulph*	

POSSESSIONS, MONEY			
Bargain, disposition to		*Puls, Sil*	Sulph
Beg, entreat, tendency to			Ars, Aur, Bell, Kali-c, Plat, Puls, Stram
Dishonesty	NAT-M	*Hyos, Lach, Tarent*	Ars
Greed, for money and possessions		*Ars, Calc, Chin, Lyc, Puls, Sep,*	Staph, Sulph
Hoarding	ARS, PULS, SIL	*Lyc, Med, Ph-ac, Sep, Sulph*	
Squandering	MERC	*Nux-v*	Bell, Calc, Caust, Stram, Sulph
Stealing, kleptomania	BELL, OP	*Absin, Art-v, Cur, Nux-v Puls, Sulph, Thuj*	Ars, Bry, Caust, Kali-c, Lyc, Sep, Sil, Staph, Stram, Sulph, Tarent

PSYCHIC SENSITIVITY			
Clairvoyance, psychic sensitivity		*Cann-i, Phos*	Acon, Anac, Arn, Calc, Carc, Hyos, Lach, Med, Sil, Stram, Tarent
Sensitive to music	NAT-C, NUX-V, SEP	*Acon, Ambr, Cham, Graph, Kreos, Lyc, Nat-m, Phos, Ph-ac, Sabin, Tarent*	Bufo, Cact, Carb-an, Caust, Coff, Dig, Merc, Nat-p, Stann, Thuj, Viol-o, Zinc
Sensitive to noise	ACON, ASAR, BELL, BOR, CHIN, CHIN-A, COFF, CON, KALI-C, NIT-AC, NUX-V, OP, SEP, SIL, THER, ZINC	*Arg-n, Ars, Aur, Bar-c, Bry, Calc, Carb-s, Carb-v, Caust, Cham, Cocc, Ferr, Fl-ac, Hell, Ign, Ip, Kali-p, Lac-c, Lach, Lyc, Lyss, Mag-m, Med, merc, Nat-c, Nat-m, Nat-s, Phos, Plat, Puls, Spig*	Alum, am-c, Ant-c, Cact, Carb-an, Cimic, Gels, Hura, Hyos, Iod, Kali-i, Mang, Mosch, Ph-ac, Rhus-t, Sabad, Stann

CHAOTIC			
Chaotic, confused behavior	CHIN, NUX-V	*Ars, Bell, Merc, Ph-ac, Phos, Puls, Staph*	
Dirtiness, un-washed appearance	SEP, STAPH, SULPH	*Plat*	Lach, Med, Merc, Nux-v, Phos
Reckless	CHAM, HEP, IGN, NIT-AC, NUX-V, SEP	*Anac, Bry, Carb-v, Cham, Kali-c, Med, Nat-m, Sul, Staph, Sulph, Zinc*	Acon, Caust, Croc, Ferr-p, Kali-i, Kali-p, Laur, Led, Nat-c, Olnd, Phos, Rheum, Stront
Untidiness	SULPH		Carc, Sil

RUSHING			
Hasty, rapid speech (See **Rapid speech**)			
Hasty, ejaculation	LYC	*Chin, Nat-m, Ph-ac, Phos, Plat, Sep, Sulph*	Calc
Hurry, always in a great hurry	ARS, BELL, MED, MERC, NAT-M, NUX-V, SULPH, TARENT	*Arg-n, Aur, Bar-c, Hyos, Lach, Puls, Stram, Thuj*	

RUSHING		
Hurry, while eating	CAUST	*Bell, Lach, Plat*	
Hurry, while walking	ARG-N, TARENT	*Sulph*	Thuj
Impatient, impetuous	CHAM, HEP, IGN, NIT-AC, NUX-V, SEP	*Anac, Bry, Carb-v, Cham, Kali-c, Med, Nat-m, Staph, Sulph, Zinc*	Acon, Caust, Croc, Ferr-p, Kali-i, Kali-p, Laur, Led, Nat-c, Olnd, Phos, Rheum, Stront
Impulsiveness, sudden desire to do something	ARG-N, IGN, PULS	*Ars, Aur, Cic, Med*	Nux-v, Rhus-t, Staph
Rapid speech	HEP, HYOS, LACH, MERC	*Bell, Camph, Ign, Mosch, Ph-ac, Sep, Stram, Thuj*	Acon, Ambr, Arn, Ars, Bry, Cann-i, Cimic, Cina, Cocc, Lyc, Lyss, Nux-v, Plb, Verat
Reckless	ARG-N, CHAM, HEP, IGN, NIT-AC, PULS, NUX-V, SEP	*Anac, Ars, Aur, Bry, Carb-v, Cham, Cic, Kali-c, Med, Nat-m, Sul, Staph, Sulph, Zinc*	Acon, Caust, Croc, Ferr-p, Kali-i, Kali-p, Laur, Led, Nat-c, Olnd, Phos, Rheum, Rhus-t, Stront
Run, tendency to, rather than walk	HYOS, STRAM	*Bell, Calc, Chin, Sulph, Tarent*	

RUSHING			
Stubbornness, obstinacy	ANAC, ARG-N, BAR-C, BELL, CALC, CHAM, NUX-V, TARENT, TUB		
Thoughts, rush of (many thoughts at once)	BELL, CANN-I, LACH, PHOS	*Ars, Ign, Kali-c, Nux-v, Ph-ac, Puls, Sil, Sulph*	Calc, Chin

HAPPINESS, SUPERFICIAL			
Cheerfulness, happiness, not related to actual events	CANN-I, HYOS, LACH		
Idealism		*Caust, Ign, Lyc, Plat*	Tub
Joking manner	IGN	*Ars, Cann-i, Hyos, Lach, Stram, Tarent*	
Laughter, immoderate	CANN-I	*Hyos, Ign, Nat-m, Plat*	Stram, Tarent
Laughter, involuntary	CANN-I, IGN	*Nat-m, Nit-ac, Tarent*	Aur, Bell, Hyos, Lyc, Phos, Puls, Sep
Laughter, over serious matters		*Anac, Nat-m, Phos*	Arg-n, Cann-i, Ign, Lyc, Plat, Sulph

HAPPINESS, SUPERFICIAL			
Laughter, during sleep	LYC, SULPH	*Hyos, Sil, Stram*	Bell, Caust, Ph-ac, Sep
Singing		*Hyos, Lach, Plat, Stram*	Bell
Singing, loud		*Hyos, Stam*	
Smiling	HYOS		Ars, Bell, Nux-v
Verses, rhymes and songs, writing or speaking in		*Cann-i, Stram*	Chin, Lach, Lyc, Nat-m, Staph

ACTIVITIES, BETTER FROM			
Dancing	SEP, TARENT (esp. break dancing)	*Ign, Nat-m*	Carc, Caust, Sil
Exercise	SEP, TARENT (esp. break dancing)	*Ign, Nat-m*	Carc, Caust, Sil
Music, (classical)	AUR		Carc, Merc, Nat-m, Thuj (church music), Tub
Music, (church music)			Thuj
Music, (fast, pop music)	TARENT		Carc, Merc, Nat-m, Tub
Music, to relieve restlessness of ex-	TARENT		

ACTIVITIES, BETTER FROM			
tremities			
Occupation, diversion, inability to watch TV without other activity	SEP	*Ign, Nux-v,*	Ars, Aur, Calc, Carc, Chin, Lach, Lyc, Puls, Sil, Stram
Company, desire for	ARG-N, ARS, HYOS, KALI-C, LAC-C, LYC, PHOS	*Apis, Calc, Camph, Clem, Con, Gels, Ign, Kali-ar, Kali-p, Lil-t, Mez, Nux-v, Pall, Puls, Sep, Stram*	Ant-t, Asaf, Aur-m, Bell, Brom, Bry, Bufo, Calc-p, Caust, Coloc, Crot-h, Dros, Hep, Merc, Plb, Tarent, Verat, Zinc

RELIGION			
Doubt of soul's welfare	ARS, AUR, LACH, PULS	*Sulph*	
Fanaticism		*Thuj*	Caust, Med, Puls, Sulph
Mania	HYOS	*Lach, Plat, Puls, Stram*	Sulph
Religious nature	HYOS, LACH, SEP, STRAM, SULPH	*Arg-n, Ars, Aur, Bell, Calc, Cham, Med, Plat, Puls*	Ign, Lyc

ENVY/JEALOUSY			
Envy, desire for what others have (not the same as jealousy	ARS	*Calc, Lach, Plat, Puls, Sep, Staph*	Lyc, Nux, v, Sulph
Jealousy, posses-siveness	HYOS, LACH, NUX-V	*Lyc, Med, Plat, Puls, Sep, Staph, Stram*	Anac, Ars, Calc, Coloc, Ign, Kali-c, Nat-m, Ph-ac

SEXUALITY			
Flirtatious	CALC, CAUST, CHIN, IGN, LYC, MED, NAT-M, NUXV, PHOS, PLAT, PULS, STRAM	*Bell, Colc, Hyos, Lach, Merc, Sep, Staph*	
Indecent dressing, stripping, streaking, nudism	HYOS	*Phos, Stram*	Bell, Merc, Tarent
Libido, low (See **Low or absent sex drive**)			
Low or absent sex drive	CAUST, GRAPH, LYC	*Agn, Bar-c, Ferr, He-lon, Hep, Mag-c, Nat-m, Ph-ac, Psor, Rhod, Sep*	Alum, Ambr, Bell, Bor, Cann-s, Carb-an, Ferr-p, Kali-i, Mur-ac, Phos, Sil, Sulph

SEXUALITY			
Lust	HYOS, LACH, LIL-T, ORIG, PHOS, PIC-AC, PLAT, STAPH	*Acon, Calc, Cann-i, Canth, Carb-v, Caust, Chin, Graph, Puls, Sel, Sep, Sil, Stram, Tar-ent*	
Nymphomania (See *Lust*)			
Satyriasis (See *Lust*)			
Sexually amorous disposition	CALC, CAUST, CHIN, IGN, LYC, MED, NAT-M, NUX-V, PHOS, PLAT, PULS, STRAM	*Bell, Colc, Hyos, Lach, Merc, Sep, Staph*	
Sexually lecherous disposition, leering		*Calc, Chin, Hyos, Lach, Lyc, Med, Nat-m, Plat, Sil*	

THOUGHTS			
Absorbed in thought	SULPH	*Arn, Nat-m, Nux-v, Puls*	
Anxious thoughts	ACON, ARG-N, ARS, ARS-I, AUR, BELL, BISM, BRY, CACT, CALC, CALC-P, CALC-S, CAMPH, CANN-I, CARB-S, CARB-V, CAUST, CON, DIG, IOD, KALI-AR, KALI-C, KALI-P, KALI-S, LYC, MEZ, NAT-A, NAT-C, NIT-AC, PHOS, PULS, RHUS-T, SEC, SULPH, VERAT	*Abrot, Acet-ac, Aeth, All-c, Alum, Ambr, Am-c, Anac, Ant-c, Ant-t, Arg-m, Arn, Asar, Bar-c, Bar-m, Bor, Bov, Canth, Carb-an, Carbo-o, Cham, Chel, Chin-a, Chin-s, Cic, Cimic, Cocc, Coc-c, Coch, Coff, Coloc, Crot-h, Cupr, Dros, Euph, Ferr, Fl-ac, Gels, Graph, Hell, Hep, Hyos, Ign, Kali-i, Kali-n, Lach, Laur, Led, Lil-t, Mag-*	

THOUGHTS			
Anxious thoughts, cont'd.		*c, Mag-m, Mag-s, Merc, Mur-ac, Nat-p, Nat-m, Nux-v, Op, Petr, Plb, Ruta, Sa-bad, Sabin, Samb, Seneg, Staph, Stram, Tab, Tar-ent, Thuj, Zinc*	
Blames self		*Acon, Ars, Aur, Dig, Hyos, Ign, Nat-m, Op, Puls, Sarr, Thuj*	Calc-p, Cob, Cycl, Hell, Hura, Lyc, Merc, Nat-a, Ph-ac
Brooding, dwelling on past disagree-able events, holding grudges	NAT-M	*Cham, Chin, Lyc, Plat, Sep, Sulph*	

THOUGHTS

| Comprehension difficult | ARG-N, BAPT, BAR-C, BAR-M, BELL, CALC, CALC-P, CALC-S, CARB-V, GRAPH, GUAJ, HELL, HYOS, KALI-BR, KALI-C, LACH, LAUR, LYC, NAT-A, NAT-C, NAT-M, NUX-M, OP, PH-AC, PHOS, PIC-AC, PLB, PULS, SENEG, SEP, SIL, STAPH, SULPH, TUB, ZINC | *Acon, Agar, Alum, Ambr, Anac, Apis, Arg-m, Bov, Cann-s, Carb-s, Caust, Cham, Chel, Chin, Chin-s, Cic, Clem, Cocc, Colch, Con, Cop, Dig, Glon, Hep, Hydr-ac, Kali-s, Kreos, Lyss, Mag-m, Meli, Merc, Merc-c, Mez, Nat-p, Nat-s, Nit-ac, Nux-v, Olnd, Petr, Psor, Rhod, Rhus-t, Sars, Sec, Spig, Spong, Stann, Stram, Tab, Tar-ent, Thuj, Verat* | Abies-n, Abrot, Acet-ac, Aesc, Aeth, Agn, Am-c, Ant-c, Arn, Ars, Ars-i, Asar, Aster, Aur, Berb, Bism, Bufo, Cact, Calc-ar, Camph, Cann-i, Canth, Caps, Carb-ac, Carbo-o, Chin-a, Cimic, Cocc-c, Crot-t, Cupr, Cupr-ar, Cycl, Dros, Dulc, Echin, Gins, Helon, Ign, Ind, Iod, Ip, Iris, Kali-i, Kali-n, Kali-p, Lac-c, Lact, Led, Lepi, Lil-t, Med, Mosch, Mur-ac, Myric, Naja, Nicc, Par, Phys, Pip-m, Plat, Ptel, Ran-b, Ran-s, Rheum, Rhus-v, Ruta, Sa-bad, Sal-ac, Sang, Still, Sul-ac, Sumb, Teucr, Ther, Til, Valer, Verb, Viol-o |

THOUGHTS			
Confused	BELL, BRY, CALC, CANN-I, CARB-V, GLON, LACH, MERC, NAT-M, NUX-M, NUX-V, ONOS, OP, PETR, RHUS-T, SEP, SIL, STRYCH	*Acon, Agar, Anac, Apoc, Arg-n, Arn, Ars, Asar, Bapt, Bar-c, Bar-m, Bor, Bufo, Calc-p, Camph, Cann-s, Canth, Carb-an, Carb-s, Chel, Chin, Coff, Coloc, Con, Croc, Crot-c, Crot-h, Cupr, Dros, Dulc, Ferr, Gels, Graph, Hell, Hy-per, Kali-c, Kali-i, Lac-c, Laur, Lyc, Mag-c, Med, Mez, Nat-c, Ph-ac, Phos, Plb, Psor, Puls, Sabad, Sec, Seneg, Spig, Staph, Stram,*	

THOUGHTS			
Confused, cont'd.		*Sulph, Tab, Thuj, Verat, Zinc*	
Difficulty thinking	ARG-N, BAPT, BAR-C, BAR-M, BELL, CALC, CALC-P, CALC-S, CARB-V, GRAPH, GUAJ, HELL, HYOS, KALI-BR, KALI-C, LACH, LAUR, LYC, NAT-A, NAT-C, NAT-M, NUX-M, OP, PH-AC, PHOS, PIC-AC, PLB, PULS, SENEG, SEP, SIL, STAPH, SULPH, TUB, ZINC	*Acon, Agar, Alum, Ambr, Anac, Apis, Arg-m, Bov, Cann-s, Carb-s, Caust, Cham, Chel, Chin, Chin-s, Cic, Clem, Cocc, Colch, Con, Cop, Dig, Glon, Hep, Hydr-ac, Kali-s, Kreos, Lyss, Mag-m, Meli, Merc, Merc-c, Mez, Nat-p, Nat-s, Nit-ac, Nux-v, Olnd, Petr, Psor, Rhod, Rhus-t, Sars, Sec, Spig, Spong, Stann,*	Abies-n, Abrot, Acet-ac, Aesc, Aeth, Agn, Am-c, Ant-c, Arn, Ars, Ars-i, Asar, Aster, Aur, Berb, Bism, Bufo, Cact, Calc-ar, Camph, Cann-i, Canth, Caps, Carb-ac, Carbo-o, Chin-a, Cimic, Cocc-c, Crot-t, Cupr, Cupr-ar, Cycl, Dros, Dulc, Echin, Gins, Helon, Ign, Ind, Iod, Ip, Iris, Kali-i, Kali-n, Kali-p, Lac-c Lact, Led, Lepi, Lil-t, Med, Mosch, Mur-ac, Myric, Naja, Nicc, Par, Phys, Pip-m, Plat, Ptel, Ran-b, Ran-s, Rheum, Rhus-v, Ruta, Sa-bad, Sal-ac, Sang, Still, Sul-ac, Sumb, Teucr, Ther, Til, Valer,

Difficulty thinking, cont'd.		*Stram, Tab, Tar-ent, Thuj, Verat*	Verb, Viol-o
Forgets what he was about to do		*Bar-c, Card-m, Chel, Nux-m, Onos, Sulph*	Agn, Bell, Calc-p, Calc-s, Carb-ac, Cann-s, Cinnb, Fl-ac, Gran, Hydr, Jug-c, Kreos, Manc
Forgets what he was about to say	HELL	*Arg-n, Arn, Cann-l, Med, Nat-m, Onos, Sulph*	Bar-c, Cann-s, Carb-an, Colch, Hydr, Hyper, Lil-t, Merc, Nux-m, Podo, Rhod, Stram, Thuj, Verat
Introspection, meditativeness	ACON, IGN, PULS, SULPH	*Aur, Chin,*	Sep
Memory, weakness for what he was about to do		*Bar-c, Card-m, Chel, Nux-m, Onos, Sulph*	Agn, Bell, Calc-p, Calc-s, Carb-ac, Cann-s, Cinnb, Fl-ac, Gran, Hydr, Jug-c, Kreos, Manc
Memory, weakness for what he was about to say	HELL	*Arg-n, Arn, Cann-l, Med, Nat-m, Onos, Sulph*	Bar-c, Cann-s, Carb-an, Colch, Hydr, Hyper, Lil-t, Merc, Nux-m, Podo, Rhod, Stram, Thuj, Verat

THOUGHTS			
Narrow-mindedness			Bar-c, Puls
Obsessive-compulsive disorder (See also One-track mind, Repetitive Actions, Repetitive Thoughts, Washing, etc.)	ARS, HYOS, IGN, SYPH, MED, NUX-V, PULS, STAPH, THUJ	*Anac, Arg-n, Bar-c, Calc, Carc Lach, Lyc, Med, Nat-m, Nat-s, Plat, Sil, Stram, Sulph, Verat*	Aur, Iod, Psor, Sulph
One-track mind	IGN, SIL	*Acon, Anan, Carb-v, Camph, Hel, Nux-m, Puls, Stram, Sulph, Thuj*	
Philosophizing, ability for and tendency to	SULPH	*Nat-c*	Anac, Coff, Hydrog, Lach, Nit-ac
Plans, making many	CHIN	*Sulph*	Anac, Arg-n, Nux-v, Sep
Rapid thoughts	BELL, COFF, HYOS, STRAM	*Acon, Agar, Cann-I, Lach, Op, Ox-ac*	Acon, Aesc, Caust, Ign, Kalm, Onos, Sabad, Verat, Viol-o
Repetitive Actions			Chen-a, Lach, Plat, Zinc
Repetitive thoughts	CANN-I, STRAM, SULPH	*Chin, Lach, Sep, Sil*	Aur, Mag-m

THOUGHTS			
Reproaches self		*Acon, Ars, Aur, Dig, Hyos, Ign, Nat-m, Op, Puls, Sarr, Thuj*	Calc-p, Cob, Cycl, Hell, Hura, Lyc, Merc, Nat-a, Ph-ac
Thinking in circles		*Cann-i*	
Thoughts, rush of (many thoughts at once)	BELL, CANN-I, LACH, PHOS	*Ars, Ign, Kali-c, Nux-v, Ph-ac, Puls, Sil, Sulph*	Calc, Chin
Tormenting thoughts	NAT-M	*Ars, Caust, Lach, Lyc, Nit-ac, Sulph.*	
Vanishing thoughts	NUX-M	*Anac, Asar, Bry, Calc, Cann-s, Gels, Lach, Lyc, Manc, Merc, Mez, Nat-m, Nit-ac, Psor, Puls*	Apis, Apoc, Bapt, Bell, Bor, Calc-s, Canth, Carb-an, Cham, Chel, Cic, Coff, Cupr, Euon, Guaj, Hell, Hep, Iod, Kali-bi, Kali-c, Kali-p, Kreos, Lac-c, Laur, Med, Nux-v, Op, Ph-ac, Ran-b, Rhod, Rhus-t, Staph, Sulph, Viol-o, Zinc

THOUGHTS			
Wandering thoughts		*Acon, Aloe, Arn, Bapt, Graph, Puls, Zinc*	Anac, Bell, Cann-I, Cann-s, Dig, Glon, Iod, Kali-br, Lyc, Merc, Nat-c, Nat-m, Op, Ph-ac, Plb, Staph, Valer

CONCERN			
Injustice, strongly affected by, cannot support the unjust	CAUST, STAPH		Calc, Ign, Med, Merc, Nat-m, Nux-v, Phos, Sulph
Inquisitiveness		*Sulph*	Aur, Hyos, Lach, Puls, Sep
Kindness, sympathy	PHOS	*Caust, Ign, Med, Nat-m, Nit-ac, Nux-v*	Carc
Sentimentality	IGN, NUX-V	*Nat-m, Phos, Puls, Sulph, Tub*	
Worry, anxiety about friends at home		*Phos, Sulph*	
Worry, anxiety about others	CAUST	*Arg-n, Ars, Carc, Nux-v, Phos, Staph, Sulph*	

INDIFFERENCE			
In general	GELS, HELL, PHOS, PH-AC	*Anac-oc, Arg-n, Calc, Cocc, Con, Kali-br, Staph*	Agn, Ant-t, Arum-t, Berb, Bism, Bufo, Camph, Chin, Crot-h, Cupr-s, Ign, Jatr, Lach, Sec, Sil
In company, in society	ARG-N,	*Plat*	Kali-c, Lyc, Nat-m
To business		*Arg-n, Arn, Ph-ac, Puls, Sep, Strain, Sulph*	
To loved ones, relatives	PHOS, SEP	*Acon,* Ars, *Plat*	Bell, Carc, Merc
To opposite sex	SEP	*Puls*	Thuj
To personal appearance	SULPH		
To welfare of others	SULPH	*Nux-v*	Ars, Caust, Lach, Nat-m, Plat

SPEECH			
Conversation (See **Talk**, **Talkative-ness**)			
Ejaculations of speech	LYC	*Chin, Nat-m, Ph-ac, Phos, Plat, Sep, Sulph*	Calc
Gossiping			Ars, Calc, Caust, Hyos, Lach, Stram
Loud voice			Bell, Cann-i, Hyos, Lach, Mosch, Nux-m, Sulph
Rapid speech	HEP, HYOS, LACH, MERC	*Bell, Camph, Ign, Mosch, Ph-ac, Sep, Stram, Thuj*	Acon, Ambr, Arn, Ars, Bry, Cann-i, Cimic, Cina, Cocc, Lyc, Lyss, Nux-v, Plb, Verat
Sighing	BRY, CALC-P, CIMIC, IGN	*Aur, Carc, Cham, Cocc, Cycl, Glon, Hell, Hydrog, Indg, Ip, Nat-m, Nat-p, Nux-v, Ph-ac, Puls, Sep, Stram*	Alum, Am-m, Anac, Arg-n, Cact, Carb-ac, Gels, Hyos, Iber, Lach, Lil-t, Lyc, Phos, Plat, Plb, Podo, Psor, Ran-s, Sil, Staph, Sulph, Thuj, Verat-v, Zinc

SPEECH			
Slow speech	HELL, LACH	*Aeth, Arg-n, Kali-br, Op, Ph-ac, Phos, Plb, Sec, Sep, Thuj*	Ars, Carb-an, Chin-s, Cocc, Merc, Morph, Nat-c, Nux-m, Petr, Phys, Rhus-t, Syph
Soft or low voice		*Arn, Ars, Cact, Calc, Canth, Ign, Spong*	Alumn, Ant-c, Cann-i, Cham, Chin, Hep, Hydrog, Lyc, Ox-ac, Puls, Sec, Sul-ac, Verat
Talk, indisposed to	AUR, CARB-AN, COCC, GLON, PH-AC, PHOS, PLAT, PULS, SULPH, VERAT, ZINC	*Acon, Agar, Ant-c, Arg-m, Arg-n, Arn, Ars, Arund, Bar-c, Bell, Calc, Caps, Carb-v, Caust, Chin, Ci-mic, Coloc, Euph, Ferr, Gels, Hell, Hep, Hipp, Hyos, Ign, Lyc, Lycps, Mag-m, Mang, Merc, Mur-ac, Nat-m, Nat-s, Nit-ac, Nux-v, Pic-ac, Plb, Rhus-t, Scorp, Stann, Staph,*	Abrot, Alco, Alum, Am-c, Am-m, Anac, Ant-t, Apoc, Aster, Bapt, Bar-m, Berb, Bor, Brom, Bry, Bufo, Cact, Calc-p, Calc-s, Cann-i, Cann-s, Canth, Carb-ac, Carb-s, Carc, Cham, Chel, Cic, Cina, Clem, Colch, Crot-t, Cupr, Cycl, Dig, Dros, Euphr, Fl-ac, Graph, Grat, Guai, Ham, Helon, Hydr, Iod, Ip, Piloc, Kali-ar, Kali-bi, Kali-c, Kali-m, Kali-p, Kali-s, Kali-Sil, Kreos, Lac-ac, Lach,

SPEECH			
Talk, indisposed to, cont'd.		*Stram, Tarent, Thuj*	Led, Lil-t, Mag-c, Mag-s, Manc, Many, Mez, Mosch, Murx, Naja, Nat-a, Nat-c, Nat-p, Nux-m, Onos, Op, Ox-ac, Petr, Phys, Sabin, Sec, Sep, Sil, Spig, Spong, Stront-c, Sul-ac, Tarax, Tub, Ust, Zinc-p
Talkativeness, or tendency to voluminous letters	HYOS, LACH, STRAM	*Agar, Arg-m, Aur, Bell, Cann-i, Cimic, Cocc, Cupr, Gels, Iod, Kali-i, Mosch, Mur-ac, Nat-c, Op, Par, Phos, Plb, Podo, Pyrog, Verat*	Abrot, Acon, Agn, Alco, Ant-t, Apis, Ars-i, Bar-c, Bar-i, Calc, Caust, Chel, Coff, Crot-h, Dulc, Glon, Hep, Lil-t, Lyc, Nat-a, Nat-m, Nux-m, Nux-v, Psor, Rhus-t, Staph, Sulph, Tarax, Tarent, Thea, Ther, Thuj, Va-ler, Verat-v, Zinc

WILLFULNESS			
Courage		*Ign, Puls, Tub*	
Optimism		*Calc, Sulph*	Lyc, Nux-v, Puls, Sil, Tub
Stubbornness, obstinacy	ANAC, ARG-N, BAR-C, BELL, CALC, CHAM, NUX-V, TARENT, TUB		
Stubbornness, obstinacy in children	CALC, TUB	*Carc, Cham, Chin*	
Work, compulsion to	AUR, NUX-V, TARENT, TUB	*Bar-c, Carc, Hyos, Ign, Lach, Lyc, Sep*	

LACK OF WILL			
Ambition, loss of		*Sep*	Arg-n, Ars, Caust
Dependent, co-dependency, etc.	AUR, HYOS, PH-AC, PHOS, STAPH	*Bell, Caust, Gels, Ign, Kali-c, Lach, Lil-t, Lyc, Nux-v, Puls, Sep, Tarent*	Arg-n, Ars, Brom, Bry, Calc, Calc-p, Camph, Clem, Con, Crot-h, Dros, Hep, Kali-p, Pall, Stram, Verat, Zinc

LACK OF WILL			
Laziness, especially about work	CHIN, LACH, NAT-M, NIT-AC, NUX-V, PHOS, PULS, SEP, SULPH, TUB		
Will, divided	ANAC	*Lach*	
Will, loss of, lack of drive		*Calc, Lyc, Merc, Nat-m, Phos*	Bar-c, Sil, Sulph
Will, weakness of	CALC	*Anac, Ars, Bar-c, Lach, Lyc, Tarent*	Caust, Chin, Ign, Kali-c, Merc, Nat-m, Nux-v, Puls, Sil, Staph, Sulph

SECTION 2.
PERSISTENT STATES

Abandoned feeling	AUR, PULS, PSOR	*Arg-n, Cycl, Lach, Plat, Stram*	Alum, Bar-c, Calc, Camph, Cann-i, Carb-an, Carb-v, Chin, Kali-br, Kali-c, Lac-d, Lil-t, Lth, Lyss, Mag-m, Nat-c, Pall, Rhus-t, Valer, Verat
Abusive		*Anac, Bell, Hyos, Lyss, Nux-v, Petr, Seneg, Sep, Verat*	Am-c, Am-m, Caust, Con, Dulc, Ip, Lyc, Mosch, Nit-ac, Plb, Ran-b, Spong, Stram, Tub, Viol-t
Alone, desire to be	ANAC, BAR-C, CARB-AN, CHAM, CIC, GELS, IGN, NAT-M, NUX-V	*Ambr, Aur, Bel, Bry, Cact, Calc-p, Carb-v, Chin, Coloc, Cupr, Cycl, Ferr, Hell, Hep, Hyos, Iod, Lach, Led, Lyc, Nat-c, Plat, Puls, Rhus-t, Sel, Sep, Stann, Sulph, Thuj*	Acon, Alum, Anan, Ant-c, Ant-t, Bufo, Calc, Cann-i, Carb-s, Cedr, Cimic, Clem, Coca, Con, Cop, Dig, Dios, Elaps, Fl-ac, Graph, Grat, Ham, Helon, Hydr, Kali-bi, Kali-br, Kali-c, Kali-p, Kali-s, Mag-m, Mang, Nat-p, Nicc, Petr, Phos, Pic-ac, Psor, Sul-ac, Tarent, Til, Verat

Angry	ANAC, ARS, AUR, BRY, CHAM, HEP, IGN, KALI-C, KALI-S, LYC, NAT-M, NIT-AC, NUX-V, PETR, SEP, STAPH, SULPH	*Acon, Apis, Ars-i, Bell, Calc, Calc-p, Calc-s, Caps, Carb-an, Carb-s, Carb-v, Caust, Cocc, Coff, Coloc, Con, Croc, Dulc, Graph, Hyos, Iod, Ip, Kali-p, Led, Mez, Mosch, Mur-ac, Nat-s, Pall, Ph-ac, Phos, Psor, Rhus-t, Stann, Stront, Tarent, Thuj, Zinc*	Agar, Agn, Am-c, Arg-m, Arg-n, Arn, Asar, Aster, Bar-c, Bar-m, Bufo, Cact, Calad, Cann-s, Canth, Chel, Chin, Chin-a, Cimicic, Clem, Cop, Crot-t, Cur, Cypr, Cycl, Dig, Dros, Elaps, Eupi, Ferr, Ferr-ar, Ferr-i, Ferr-p, Fl-ac, Gels, Gran, Ham, Hell, Hydr, Kali-ar, Kali-n, Lach, Mag-s, Mang, Meph, Merc, Nat-a, Nat-c, Nat-p, Nicc, Nux-m, Olnd, Op, Osm, Plat, Puls, Ran-b, Rat, Ruta, Sabad, Sang, Seneg, Sil, Spig, Squil, Stram, Sul-ac, Tell, Trill, Valer, Verat
Anxiety (See **Anxious**)			

Anxious	ACON, ARG-N, ARS, ARS-I, AUR, BELL, BISM, BRY, CACT, CALC, CALC-P, CALC-S, CAMPH, CANN-I, CARB-S, CARB-V, CAUST, CON, DIG, IOD, KALI-AR, KALI-C, KALI-P, KALI-S, LYC, MEZ, NAT-A, NAT-C, NIT-AC, PHOS, PULS, RHUS-T, SEC, SULPH, VERAT	*Abrot, Acet-ac, Aeth, All-c, Alum, Ambr, Am-c, Anac, Ant-c, Ant-t, Arg-m, Arn, Asar, Bar-c, Bar-m, Bor, Bov, Canth, Carb-an, Carbo-o, Cham, Chel, Chin-a, Chin-s, Cic, Cimic, Cocc, Cocc, Coch, Coff, Coloc, Crot-h, Cupr, Dros, Euph, Ferr, Fl-ac, Gels, Graph, Hell, Hep, Hyos, Ign, Kali-i, Kali-n, Lach, Laur, Led, Lil-t, Mag-c, Mag-m, Mag-s, Merc, Mur-ac, Nat-p, Nat-m, Nux-v, Op, Petr, Plb, Ruta,*	

Anxious, cont'd.		*Sabad, Sabin, Samb, Seneg, Staph, Stram, Tab, Tar-ent, Thuj, Zinc*	
Anxious with sense of suf-focation	ACON	*Ars, Phos, Staph, Stram, Sulph*	Lach, Merc, Nux-v
Argumenta-tive (See **Quarrel-some**)			

Attention Deficit / Hyperactivity	**(Begin with the following pool, then individualize with other characteris-tics of Pt.)**		
	Listed in order of broad ac-tion:		
	1. Lyc		
	2. Nux-v		
	3. Bell		
	4. Anac		
	5. Stram		
	6. Hyos		
	7. Verat		
	8. Merc		
	9. Bar-c		
	10. Calc		
	11. Lach		
	12. Cham		
	13. Phos		
	14. Sulph		
	15. Ars		
	16. Plat		
	17. Nat-m		
	18. Nit-ac		
	19. Aur		
	20. Caust		
	21. Sep		
	22. Sil		
	23. Hep		
	24. Petrol		

Attention Deficit / Hyperactivity, cont'd.	25. Puls 26. Apis 27. Arg-n 28. Canth 29. Carbo-v 30. Carb-s 31. Kali-c 32. Lac-c 33. Zinc 34. Cupr 35. Calc-p 36. Med 37. Tub 38. Lyss 39. Cina 40. Cic **Also of use:** Acon, Aeth, Agar, Ars-i, Brom, Bufo, Can-i, Gall-ac, Hell, Manc, Op, Tarent Restlessness/ Inattention: 1. Lyc 2. Med 3. Puls 4. Sulph 5. Tub 6. Tarent 7. Arg-n 8. Veratr 9. Lach 10. Ars		

Attention Deficit / Hyperactivity, cont'd.	**Anger / Aggression:** 1. Cham 2. Cina 3. Hep 4. Nux-v 5. Aur 6. Stram 7. Hyos 8. Lyc 9. Tub **Scholastic problems:** 1. Gels 2. Calc 3. Bar-c 4. Agar 5. Anac 6. Aeth 7. Sil 8. Bufo **Stubbornness:** 1. Anac 2. Arg 3. Ars 4. Calc 5. Cham 6. Nux-v 7. Tub 8. Sulph 9. Tarent		

Attention Deficit / Hyperactivity, cont'd.	Jealousy: 1. Lach 2. Plat 3. Lyc Fear of dark or being alone at night: • Stram • Brom • Med • Calc • Phos • Puls • Caust • Arg-n Tantrums: • Tub • Cham • Stram Unable to retain information: • Lyc • Anac • Med A.D.D. after grief (from death, divorce, disappointment): • Ign		

Attention Deficit / Hyperactivity, cont'd.	Low self-esteem: • Nat-m • Stram • Aur • Anac Check food cravings to further individualize.		
Blames self	ARS, AUR, HYOS, IGN, NAT-M, PULS, SULPH	*Lach, Lyc, Med, Merc, Sil, Stram, Thuj*	Calc-p, Cob, Cycl, Hell, Hura, Lyc, Merc, Nat-a, Ph-ac
Busy, must be	AUR, TARENT	*Bar-c, Croc, Hyos, Ign, Lach, Op, Sep*	Agar, Arn, Ars, Bell, Bry, Calc, Calc-p, Caps, Chin, Clem, Dig, Indg, Ip, Kreos, Led, Mag-c, Mez, Mosch, Mur-ac, Nat-c, Nat-s, Phos, Pip-m, Plb, Rhus-t, Stann, Sul-ac, Verat
Changeable, vacillating	ACON, ALUM, AMBRA, ASAF, CIMIC, CROC, IGN, MOSCH, NUX-M, PLAT, PULS, VAL, ZINC-V	*Caust, Coff, Gels, Kali-p, Lil-t, Mang-ac, Nat-m, Phos, Sep, Tarent*	

Confidence (low)	ANAC, SIL	*Aur, Bry, Chin, Kali-c, Lyc, Med, Nat-m, Ph-ac, Puls*	Am-c, Alum, Bar-c, Calc, Chin, Nit-ac, Phos, Staph
Conversation, disinclined to (See **Talk, indisposed to**)			
Critical	ARS, SULPH	Arn, Bar-c, Caust, Sulph, Ip, Lach, Lyc, Mez, Nux-v, Plat, Sep, Verat	Cham, Graph, Merc, Nat-m, Staph
Dependent, co-dependency, etc.	AUR, HYOS, PH-AC, PHOS, STAPH	*Bell, Caust, Gels, Ign, Kali-c, Lach, Lil-t, Lyc, Nux-v, Puls, Sep, Tarent*	Arg-n, Ars, Brom, Bry, Calc, Calc-p, Camph, Clem, Con, Crot-h, Dros, Hep, Kali-p, Pall, Stram, Verat, Zinc
Depressed	ACON, ARS, ARS-I, AUR, CALC, CALC-AR, CARBO-AN, CARB-S, CARC, CAUST, CHIN, FERR, FERR-I, GRAPH, HELL, IOD, IGN, KALI-BR, KALI-P,	*Bar-c, Bry, Calc-fl, Calc-p, Chin-s, Con, Croc, Cupr, Ferr-p, Helon, Hyos, Kali-c, Naja, Nat-p, Nux-v, Petr, Ph-ac, Phos, Phyt, Plb,*	

Depressed, cont'd.	LAC-C, LACH, MERC, MEZ, MURX, NAT-C, NAT-M, NAT-S, NIT-AC, PLAT, RHUS-T, PSOR, PULS, SEP, SULPH, THUJ, VERAT, ZINC	*Ruta, Sil, Spig, Spong, Staph, Still, Stram, Sul-ac, Tab, Verat-v*	
Discontented	ANAC, CALC-P, MERC, NAT-M, SULPH	*Am-m, Ars, Aur, Bism, Bor, Bry, Cham, Chel, Chin, Cina, Colch, Cupr, Hep, Kali-c, Lyc, Nit-ac, Nux-v, Pall, Plat, Puls, Rhus-t, Sep, Sil, Stann, Staph, Thuj*	Acon, Aeth, Agar, Agn, Alet, Aloe, Alum, Am-c, Apis, Arn, Ars-i, Asar, Aur-m, Bar-c, Bell, Berb, Bov, Brom, Calc, Calc-s, Cann-s, Canth, Caps, Carbo-an, Carb-s, Caust, Chin-a, Cic, Clem, Cocc, Coff, Coloc, Con, Crot-t, Dulc, Ferr, Ferr-p, Fl-ac, Graph, Grat, Ham, Hell, Hura, Ign, Indg, Iod, Ip, Kreos, Lach, Laur, Led, Lil-t, Mag-c, Mag-m, Mag-s, Manc, Mang, Mez, Mur-ac, Nat-a, Nat-c, Nat-p,

Discontented, cont'd.			Nit-ac, Op, Orig, Par, Petr, Ph-ac, phos, Plb, Ran-b, Rhod, Rob, Ruta, Samb, Sars, Spong, Tarent, Ther, Til, Viol-t
Domineering	NAT-M, SEP	*Camph, Lyc, Merc*	Anac, Aur, Caust, Cham, Con, Cupr, Ferr, Hep, Lach, Plat, Sulph, Verat
Fastidious	ARS, NUX-V,	*Anac, Carc, Graph, Nat-m, Nat-s, Lyc, Puls, Sil*	
Fearful	ACON, ARG-N, AUR, BELL, BOR, CALC, CALC-P, CARB-S, CARC, CIC, DIG, GRAPH, IGN, KALI-AR, LYC, LYSS, NAT-C, PHOS, PLAT, PSOR, SEP, STRAM, ZINC-P.	*Ars, Bar-c, Bry, Cact, Calc-s, Caps, Carb-v, Caust, Cocc, Con, Gels, Hep, Hyos, Iod, Kali-br, Kali-i, Mag-m, Merc, Mosch, Nat-m, Nat-p, Nux-v, Petr, Phyt, Puls, Rhus-t, Spong, Sulph, Sul-*	

Fearful, cont'd.		*ac, Tab, Verat, Zinc*	
Grieving	AUR, CARC, CAUST, IGN, NAT-M, PULS	*Caust, Cocc, Coloc, Graph, Lach, Lyc, Merc, Nux-v, Ph-ac, Phos, Staph, Tarent*	Acet-ac, Agar, Alum, Am-c, Am-m, Bar-c, Carb-an, Hyos, Sep, Sul-ac, Verat
Guilt-ridden	ARS, AUR, SULPH	*Acon, Bell, Carc, Caust Chel, Dig, Hyos, Ign, Lach, Med, Merc, Nat-m, Nat-m, Nux-v, Op, Ph-ac, Psor, Puls, Sil, Staph, Stram, Thuj*	Alum-m, Calc-p, Hell, Lyc, Merc, Nat-a
Hypervigi-lance, Post-traumatic Stress Disor-der (PTSD), etc.	ARN, ARS, STRAM	*Arg-n, Ign, Lach*	Arn, Aur, Bell, Cann-i, Chin, Hyos, Kali-c, Thuj
Indifferent	APIS, CARB-V, CHIN, CROT-C, HELL, LIL-T, MEZ, NAT-C, NAT-M, NAT-P, ONOS, OP, PH-AC, PHOS, PLAT,	*Agar, Alum-m, Anac, Bar-c, Carb-an, Calc, Gels, Graph, Kali-c, Kali-p, Nit-ac, Nux-v,*	Bar-m, Ferr, Stram, Sulph, Tarent, Zinc

309

Indifferent, cont'd.	SEP, STAPH	*Phos, Puls, Psor, Sep, Sil, Staph, Thuj, Verat*	
Industrious	AUR, TARENT	*Bar-c, Croc, Hyos, Ign, Lach, Op, Sep*	Agar, Arn, Ars, Bell, Bry, Calc, Calc-p, Caps, Chin, Clem, Dig, Indg, Ip, Kreos, Led, Mag-c, Mez, Mosch, Mur-ac, Nat-c, Nat-s, Phos, Pip-m, Plb, Rhus-t, Stann, Sul-ac, Verat
Irritable	ALUM, ANT-C, APIS, AUR, BELL, BOV, BRY, CALC, CALC-S, CARB-V, CAUST, CHAM, GRAPH, HEP, KALI-C, KALI-I, KALI-S, LIL-T, LYC, MAG-C, NAT-C, NAT-M, NAT-C, NAT-M, NIT-AC, NUX-V, PETR, PH-AC, PHOS, PLAT, PULS, RHUS-T, RAN-B, SEP, SIL, STAPH, STRY, SULPH,	*Ars, Bar-c, Bor, Calc-p, Caul, Chin, Clem, Coff, Con, Croc, Crot-h, Cycl, Dig, Dulc, Ferr, Gels, He-lon, Iod, Ip, Kali-bi, Kali-p, Lach, Led, Mang, Med, Merc, Mur-ac, Murx, Nat-s, Psor, Spig, Spong, Stram, Tarent, Verat*	

Irritable, cont'd.	SUL-AC, THUJ, VERAT-V, ZINC		
Jealous	HYOS, LACH	*Apis, Lyc, Med, Nux-v, Plat, Puls, Sep, Staph, Stram*	Anac, Ars, Calc, Caust, Coloc, Ign, Kali-c, Nat-m, Ph-ac
Lazy	CARB-S, CHEL, CHIN, GRAPH, LACH, LYC, NAT-M, NIT-AC, NUX-V, PHOS, PIC-AC, PULS, SULPH, TUB,	*Arg-n, Aur, Calc, Calc-p, Carb-v, Hep, Iod, Lac-c, Lyc, Mag-m, Ph-ac, Phos, Pic-ac, Psor, Puls, Sa-bin, Thuj, Zinc*	
Loquacity (See **Talka-tive**)			
Lies, tries to appear truth-ful		*Op, Verat*	Alco
Lying, un-truthful		*Op, Verat*	Alco
Memory (Poor)	AMBR, ARG-N, ARS, BAR-C, CARB-S, CAUST, COCC, COLCH, CON, GLON, HELL, HEP, HYOS, KALI-P, LACH,	*Agn, Alum, Anac, Apis, Arn, Aur, Bov, Bry, Carb-an, Chin, Dig, Fl-ac, Form, Gels, Graph, He-*	

Memory (Poor), cont'd.	LYC, MED, MERC, PH-AC, PHOS, PLAT, PLB, SEP, VERAT	*Ion, Hydr, Ign, kali-br, Lac-ac, Lac-c, Laur, Mez, Nat-a, Nat-c, Nat-m, Nat-p, Nux-v, Op, Petr, Pic-ac, Puls, Sel, Spig, Stann, Staph, Sulph, Sul-ac, Tarent, Thuj, Tub, Viol-o, Zinc*	
Obsessive	IGN, SIL	*Acon, Anac, Arg-n, Ars, Calc, Carb-v, Nux-v, Carc, Hyos, Med, Nat-m, Nat-s, Plat, Puls, Staph, Verat*	Camph, Hell, Nux-m, Stram, Sulph, Thuj
Obstinate	ALUM, ANAC, ARG-N, BELL, CALC, CHAM, NUX-V, TARENT, TUB,	*Acon, Agar, Ars, Bry, Cina, Hep, Ign, Kali-c, Kali-p, Mag-m, Nit-ac, Pall, Ph-ac, Psor, Sil, Spong, Sulph, Thuj*	

Passive	ARS, BAR-C, ICALC, CALC-S, GELS, KALI-C, LYC, NAT-C, PHOS, SEP, SULPH	*Acon, Alum, Aur, Bor, Carb-s, Carb-v, Caust, Chin, Con, Cupr, Ign, Kali-ar, Kali-s, Merc, Nat-a, Nat-m, Nux-v, Puls, Ruhs-t, Sil, Spong, Staph, Stram*	
Perfectionist	IGN, SIL	*Ars, Bar-c, Lyc, Mur-ac, Nat-c, Nux-v, Stram, Sulph, Thuj*	Apis, Aur, Bry, Carb-s, Cham, Chin, Cycl, Ferr, Ferr-ar, Ferr-i, Graph, Hep, Hyos, Iod, Lac-d, Mez, Nat-a, Puls, Sec, Sep, Spig, Verat
Pessimistic	ARS, AUR, NIT-AC, NUX-V, PSOR		
Pleasure, things do not bring	ANAC, CALC-P, MERC, NAT-M, SULPH	*Am-m, Ars, Aur, Bism, Bor, Bry, Cham, Chel, Chin, Cina, Colch, Cupr, Hep, Kali-c, Lyc, Nit-ac,*	Acon, Aeth, Agar, Agn, Alet, Aloe, Alum, Am-c, Apis, Arn, Ars-i, Asar, Aur-m, Bar-c, Bell, Berb, Bov, Brom, Calc, Calc-s, Cann-

Pleasure, things do not bring, cont'd.		*Nux-v, Pall, Plat, Puls, Rhus-t, Sep, Sil, Stann, Staph, Thuj*	s, Canth, Caps, Carbo-an, Carb-s, Caust, Chin-a, Cic, Clem, Cocc, Coff, Coloc, Con, Crot-t, Dulc, Ferr, Ferr-p, Fl-ac, Graph, Grat, Ham, Hell, Hura, Ign, Indg, Iod, Ip, Kreos, Lach, Laur, Led, Lil-t, Mag-c, Mag-m, Mag-s, Manc, Mang, Mez, Mur-ac, Nat-a, Nat-c, Nat-p, Nit-ac, Op, Orig, Par, Petr, Ph-ac, phos, Plb, Ran-b, Rhod, Rob, Ruta, Samb, Sars, Spong, Tarent, Ther, Til, Viol-t
Quarrelsome	AUR, IGN, NUX-V, PETR, SULPH, TARENT,	*Acon, Anac, Bell, Bov, Brom, Bry, Camph, Caust, Cham, Con, Croc, Cupr, Dulc, Hyos, Lach, Lyc, Merc, Mosch,*	

Quarrelsome, cont'd.		*Nat-c, Nat-m, Nit-ac, Ph-ac, Phos, Plat, Psor, Ran-b, Sep, Staph, Stram, Tab, Thuj, Verat, Verat-v*	
Rage	AGAR, BELL, CANTH, HYOS, LAC-C, LYC, MOSCH, OP, STRAM, VERAT	*Acon, Anac, Ars, Camph, Carb-s, Colch, Cupr, Hell, Lach, Merc, Nat-m, Nit-ac, Phos, Puls, Sec, Sulph, Tab*	
Restless	ACON, ANAC, ARG-N, ARS, BAPT, CALC-P, CAMPH, CARC, CIMIC, COLOC, CUPR, FERR, LYC, MED, MERC, RHUS-T, SEC, SEP, SIL, STAPH, STRAM, SULPH, TARENT, ZINC	*Agar, Ant-t, Apis, Arg-m, Aur, Bov, Cann-s, Carb-s, Carb-v, Caust, Cham, Chel, Coff, Dig, Dulc, Ferr-i, Graph, Hell, Ign, Iod, Kali-br, Kali-c, Kali-n, kali-p, Kali-s, Lac-c, Lach, Led, Lil-t,*	

Restless, cont'd.		*Mang, Med, Merc-c, Mosch, Nat-a, Nat-c, Nat-m, Nit-ac, Nux-v, Op, Ph-ac, Plat, Psor, Rumx, Ruta, Samb, Stann, Tab, Tell, Thuj, Valer*	
Sensitive (Hyper)	ARG-N, BELL, BOR, CHIN, COFF, GELS, IGN, LYC, LYSS, NAT-M, NIT-AC, NUX-V, PHOS, PLB, PULS, RAN-B, SIL, SULPH, THER, VALER	*Asar, Aur, Bar-c, Bov, Calc, Caust, Cham, Cocc, Coff, Hyos, Iod, Lash, Lil-t, Nat-c, Nat-m, Nat-p, Plat, Sabin, Sep, Sil, Staph, Zinc*	
Startles easily (See *Hypervigilant*)			

Suicidal	AUR, AUR-M, NAT-S	*Anac, Ant-c, Ant-t, Ars, Calc, Caps, Chin, Cimic, Hep, Hyos, Kali-br, Lac-d, Lach, Merc, Nux-v, Plb, Psor, Puls, Sep, Spig, Stram, Zinc*	
Suspicious	ACON, ANAC, ARS, BARYTA, CARB, BRY, CAUST, LACH, LYC, PULS, RHUS-T, STRAM		
Sympathetic	CALC-P, CARC, CAUST, COCC, IGN, NAT-C, NAT-M, NIT-AC, NUX-V, PHOS, PULS		

| Talk, indisposed to | AUR, CARB-AN, COCC, GLON, PH-AC, PHOS, PLAT, PULS, SULPH, VERAT, ZINC | *Acon, Agar, Ant-c, Arg-m, Arg-n, Arn, Ars, Arund, Bar-c, Bell, Calc, Caps, Carb-v, Caust, Chin, Cimic, Coloc, Euph, Ferr, Gels, Hell, Hep, Hipp, Hyos, Ign, Lyc, Lycps, Mag-m, Mang, Merc, Mur-ac, Nat-m, Nat-s, Nit-ac, Nux-v, Pic-ac, Plb, Rhus-t, Scorp, Stann, Staph, Stram, Tarent, Thuj* | Abrot, Alco, Alum, Am-c, Am-m, Anac, Ant-t, Apoc, Aster, Bapt, Bar-m, Berb, Bor, Brom, Bry, Bufo, Cact, Calc-p, Calc-s, Cann-i, Cann-s, Canth, Carb-ac, Carb-s, Carc, Cham, Chel, Cic, Cina, Clem, Colch, Crot-t, Cupr, Cycl, Dig, Dros, Euphr, Fl-ac, Graph, Grat, Guai, Ham, Helon, Hydr, Iod, Ip, Piloc, Kali-ar, Kali-bi, Kali-c, Kali-m, Kali-p, Kali-s, Kali-Sil, Kreos, Lac-ac, Lach, Led, Lil-t, Mag-c, Mag-S, Manc, Many, Mez, Mosch, Murx, Naja, Nat-a, Nat-c, Nat-P, Nux-m, Onos, Op, Ox-ac, Petr, Phys, Sabin, Sec, Sep, Sil, Spig, Spong, Stront-c, Sul-ac, Ta-rax, Tub, Ust, Zinc-p |

Talkative-ness, or tendency to voluminous letters	HYOS, LACH, STRAM	*Agar, Arg-m, Aur, Bell, Cann-i, Cimic, Cocc, Cupr, Gels, Iod, Kali-i, Mosch, Mur-ac, Nat-c, Op, Par, Phos, Plb, Podo, Pyrog, Verat*	Abrot, Acon, Agn, Alco, Ant-t, Apis, Ars-i, Bar-c, Bar-i, Calc, Caust, Chel, Coff, Crot-h, Dulc, Glon, Hep, Lil-t, Lyc, Nat-a, Nat-m, Nux-m, Nux-v, Psor, Rhus-t, Staph, Sulph, Tarax, Tarent, Thea, Ther, Thuj, Va-ler, Verat-v, Zinc
Timid	ARS, BAR-C, ICALC, CALC-S, GELS, KALI-C, LYC, NAT-C, PHOS, SEP, SULPH	*Acon, Alum, Aur, Bor, Carb-s, Carb-v, Caust, Chin, Con, Cupr, Ign, Kali-ar, Kali-s, Merc, Nat-a, Nat-m, Nux-v, Puls, Ruhs-t, Sil, Spong, Staph, Stram*	
Violent	ANAC, AUR, BELL, CIC, HYOS, NUX-V, LACH, NAT-M, PETR, STRAM, TUB		

Vengeful	ANAC, ARS, NAT-M, NUX-V, STRAM, TUB	*Acon, Aur, Bell, Calc, Cham, Hyos, Lach, Lyc, Nit-ac, Ph-ac, Staph*	
Weary	ARS, AUR, CARC, CHIN, KALI P, NAT-M, NAT-S, MERC, NIT-AC, NUX-V, PH-AC, PHOS		
Worrying	ARS, CALC, CAUST, CHIN, IGN, LYC, NAT-C, NAT- M, PH-AC, PULS, STAPH		

SECTION 3.

STRESS TRAUMA—Causative factors

Look for the causative factor in the particular phase you are treating in a patient. Choose the one that is the strongest in the case and particularly *if it represents a deviation from the normal.* For example, jealousy and anger may figure equally for two patients, but for one, the jealousy is not part of her usual personality. The jealousy she feels is uncharacteristic and therefore more a part of her pathology. The other patient, being a jealous type in the first place, will need the anger weighed more heavily in working up her case.

Abandonment	AUR, PULS, PSOR	*Arg-n, Cycl, Lach, Plat, Stram*	Alum, Bar-c, Calc, Camph, Cann-i, Carb-an, Carb-v, Chin, Kali-br, Kali-c, Lac-d, Lil-t, Lth, Lyss, Mag-m, Nat-c, Pall, Rhus-t, Valer, Verat
Anger, suppressed	LYC, STAPH	*Aur, Ign, Nat-m*	Carc, Cham, Sep
Anger with indignation	COLOC, STAPH	*Aur, Nux-v*	Ars, Lyc, Merc, Nat-m, Plat
Anger with silent grief	IGN, LYC, STAPH	*Acon, Chin, Coloc, Nat-m, Ph-ac*	Bell, Carc, Cham, Nux-v, Plat, Puls
Anniversary reactions (one year since trauma, etc.)	ACON, IGN, NAT-M, NUX-V		

Anticipation anxiety, fear and apprehension before any event	ARG-N, ARS, CALC, CARC, IGN, GELS, LYC, MED, PHOS, PULS, SIL	*Anac, Bar-c, Caust, Ph-ac*	Nat-m
Anxiety	ACON, ARG-N, ARS, ARS-I, AUR, BELL, BISM, BRY, CACT, CALC, CALC-P, CALC-S, CAMPH, CANN-I, CARB-S, CARB-V, CAUST, CON, DIG, IOD, KALI-AR, KALI-C, KALI-P, KALI-S, LYC, MEZ, NAT-A, NAT-C, NIT-AC, PHOS, PULS, RHUS-T, SEC, SULPH, VERAT	*Abrot, Acet-ac, Aeth, All-c, Alum, Ambr, Am-c, Anac, Ant-c, Ant-t, Arg-m, Arn, Asar, Bar-c, Bar-m, Bor, Bov, Canth, Carb-an, Carbo-o, Cham, Chel, Chin-a, Chin-s, Cic, Cimic, Cocc, Coc-c, Coch, Coff, Coloc, Crot-h, Cupr, Dros, Euph, Ferr, Fl-ac, Gels, Graph, Hell, Hep, Hyos, Ign, Kali-i, Kali-n, Lach, Laur, Led, Lil-t, Mag-c, Mag-m, Mag-s, Merc, Mur-ac, Nat-p, Nat-m, Nux-v, Op, Petr, Plb, Ruta, Sabad, Sa-bin, Samb, Seneg,*	

Anxiety, cont'd.		*Staph, Stram, Tab, Tarent, Thuj, Zinc*	
Bad news, disappoint-ment	BRY, CAUST, CHAM, COCC, COLCH, COLOC, GELS, NUX-V, STAPH	*Acon, Apis, Ars, Aur, Grat, Hyos, Ign, Lach, Nat-m, Ph-ac, Puls, Sep*	
Childbirth (pu-erperal depression)	CIMIC, NAT-M, PLAT, SEP, STAPH	*Acon, Arn, Ign*	
Claustrophobia or Closed-in places	ACON, ARG-N, LYC, NAT-M, PULS, STRAM	*Calc, Ign, Med*	Ambr, Ambr, Aran, Chin-a, Cocc, kali-ar, Nit-ac, Nux-v, Plb, Psor, Ruta, Staph, Succ, Sulph, Tab, Valer
Contradiction, being told off, traumatized by		*Aur, Cham, Ign*	Anac, Med, Sil
Exams, men-tally paralyzed from	ARG-N, ARS, CALC, CARC, IGN, GELS, LYC, MED, PHOS, PULS, SIL	*Anac, Bar-c, Caust, Ph-ac*	Nat-m
Homesickness (See **Nostal-gia**)			

Humiliation, shame, being put down forcibly, cannot bear it	CHAM, COLOC, IGN, LYC, NAT-M, PH-AC, STAPH	*Acon, Anac, Arg-n, Ars, Aur, Nux-v, Puls, Sep, Sil, Stram, Sulph*	Bell, Calc, Lach, Merc, Plat
Humiliation with anger	COLOC		
Humiliation with indigna- tion	STAPH		
Menopause	CACT, CAUST, IOD, LIL-T, PULS, SEP, THUJ		
Nostalgia (homesick- ness)	CAPS, IGN, PH-AC, GELS	*Eup-purp, Hell, Mag-m, Senec*	
Noise	ACON, ASAR, BELL, BOR, CHIN, CHIN-A, COFF, CON, KALI-C, NIT-AC, NUX-V, OP, SEP, SIL, THER, ZINC	*Arg-n, Ars, Aur, Bar-c, Bry, Calc, Carb-s, Carb-v, Caust, Cham, Cocc, Ferr, Fl-ac, Hell, Ign, Ip, Kali-p, Lac-c, Lach, Lyc, Lyss, Mag-m, Med, Merc, Nat-c, Nat-m, Nat-s, Phos, Plat, Puls, Spig*	Alum, am-c, Ant-c, Cact, Carb-an, Ci-mic, Gels, Hura, Hyos, Iod, Kali-i, Mang, Mosch, Ph-ac, Rhus-t, Sabad, Stann

Offense, taking	ARS, CALC, CARC, CAUST, NUX-V, STAPH, TUB	*Acon, Aur, Bell, Coloc, Lach, Med, Nat-m, Nit-ac, Plat, Puls, Sep, Sulph, Thuj*	Sil
Offenses from the past		*Ars, Cham, Ign, Staph*	Calc
Punishment, becomes ill from punish-ment (typically children)		*Carc, Ign, Staph*	Nat-m, Tarent
Rape	STRAM		Aur, Sil, Tar-ent
Reproached, censured with severe lan-guage (See **Humiliation**)	NAT-M	*Ign, Med, Staph, Stram*	Bell, Carc, Coloc, Nux-v, Ph-ac, Plat, Tarent
Rudeness of others	STAPH	*Calc*	Anac, Carc, Med, Nat-m, Nux-v, Ph-ac
Scorn, being scorned, treated with extreme con-tempt	CHAM, NUX-V	*Aur, Coloc, Nat-m, Phos, Plat, Sep, Staph*	Acon, Hyos, Lyc, Sulph, Tarent
Sexual abuse	STAPH, STRAM		Aur, Sil, Tar-ent
Stage fright, mentally para-lyzed from	ARG-N, ARS, CALC, CARC, IGN, GELS, LYC, MED, PHOS, PULS, SIL	*Anac, Bar-c, Caust, Ph-ac,*	Nat-m
Violence	STRAM		Aur, Sil, Tar-ent

GRIEF, LOSS, & BETRAYAL Causative factors			
Abandoned	AUR, PULS, PSOR	*Arg-n, Cycl, Lach, Plat, Stram*	Alum, Bar-c, Calc, Camph, Cann-i, Carb-an, Carb-v, Chin, Kali-br, Kali-c, Lac-d, Lil-t, Lth, Lyss, Mag-m, Nat-c, Pall, Rhus-t, Valer, Verat
Betrayal	AUR, LYC, NAT-M	*Nux-v*	Ign, Lach, Merc, Ph-ac, Puls, Sep
Betrayal of ambition		*Nux-v*	Bell, Merc, Plat, Puls
Betrayal of confidence	ARG-N, AUR, NAT-C, PSOR, PULS	*Staph*	
Betrayal of friendship	ARG-N, AUR, NAT-C, PSOR, PULS	*Staph*	Ign, Nux-v, Ph-ac, Sil, Sulph
Death of a child	CAUST, IGN		Calc, Lach, Nat-m, Nux-v, Ph-ac, Plat, Staph, Sulph
Death of parents or friends	CAUST, IGN		Ars, Calc, Nit-ac, Nux-v, Plat, Staph
Death of spouse	CAUST, IGN		Ars, Calc, Nit-ac, Nux-v, Plat, Staph

GRIEF, LOSS, & BETRAYAL Causative factors			
Disappointment	AUR, CAUST, IGN, LYC, NAT-M, PH-AC, PULS, STAPH	*Cham, Coloc, Lach, Merc, Nux-v,*	Acon, Ars, *Coloc,* Hyos, Plat, Sep
Grief	AUR, CAUST, IGN, LACH, NAT-M, PHAC, PHOS, SEP	*Bell, Calc, Carc, Coloc, Hyos, Nux-v, Plat, Puls,*	Anac, Sil, Tub
Grief, but can-not cry	NAT-M	*Ign*	Carc, Nux-v, Puls
Grief, sudden			Ign
Homesickness	PH-AC	*Gels*	Ign
Love, disap-pointed	AUR, HYOS, NAT-M, PH-AC, STAPH	*Bell, Caust, Lach*	Kali-c, Nux-v, Phos, Sep, Sulph, Tarent
Love, with si-lent grief	IGN, NAT-M, PH-AC		Phos

FEAR — Causative factors			
Anticipation anxiety, stage fright, nervous before trips, appointments, exams, etc. (See also *Public, appearing in*)	ARG-N, ARS, CALC, CARC, IGN, GELS, LYC, MED, PHOS, PULS, SIL	*Anac, Bar-c, Caust, Ph-ac,*	Nat-m
Company, desire to be alone	ANAC, BAR-C, CARB-AN, CHAM, CIC, GELS, IGN, NAT-M, NUX-V	*Ambr, Aur, Bel, Bry, Cact, Calc-p, Carb-v, Chin, Coloc, Cupr, Cycl, Ferr, Hell, Hep, Hyos, Iod, Lach, Led, Lyc, Nat-c, Plat, Puls, Rhus-t, Sel, Sep, Stann, Sulph, Thuj*	Acon, Alum, Anan, Ant-c, Ant-t, Bufo, Calc, Cann-i, Carb-s, Cedr, Cimic, Clem, Coca, Con, Cop, Dig, Dios, Elaps, Fl-ac, Graph, Grat, Ham, Helon, Hydr, Kali-bi, Kali-br, Kali-c, Kali-p, Kali-s, Mag-m, Mang, Nat-p, Nicc, Petr, Phos, Pic-ac, Psor, Sul-ac, Tarent, Til, Verat
Embarrassment	SULPH	*Ign*	Coloc, Ph-ac, Plat, Sep, Staph
Fear	ACON	*Caust, Ign, Sil.*	Arg-n, Bell, Calc, Carc, Lyc, Phos, Puls

FEAR — Causative factors			
Fright, startled	ACON, CAUST, IGN, LYC, NAT-M, PH-AC, PHOS, PULS	*Arg-n, Aur, Bell, Calc, Hyos, Lach, Nux-v, Plat, Sep, Stram*	
Fright from sight of an accident	ACON		
Public, appearing in	ARG-N, GELS, LYC, SIL	*Carb-v*	Anac, Plb

MISCELLANEOUS Causative factors			
Alcoholism		*Alco, Carc, Lach, Med, Nux-v, Op, Thuj*	Ars, Calc, Lach, Ph-ac, Sulph
Bad news	CALC	*Arn, Cham, Coloc, Ign, Med, Merc, Nat-m, Nux-v, Staph, Sulph*	
Business failure			Calc, Coloc, Nat-m, Nux-v, Ph-ac, Puls, Sep, Sulph
Discords between superior and subordinates			Lach, Merc, Nat-m, Nit-ac, Nux-v, Sulph

MISCELLANEOUS Causative factors			
Discord between parents, friends		*Nat-m*	Ars, Lach, Merc, Nit-ac, Nux-v, Sulph
Domination by others, long history of			Carc, Lyc, Sep
Excitement, emotional	PH-AC, PULS, STAPH, TUB	*Acon, Arg-n, Aur, Bell, Calc, Caust Nat-m, Nux-v, Phos, Tarent*	Sep
Excitement, sexual		*Nat-m, Plat*	Staph
Horror, ghost stories, sad stories,	CALC, PHOS, STAPH	*Aur, Caust, Lach, Lyc, Nit-ac, Nux-v, Puls, Sep, Sil, Sulph*	
Hurry		*Acon, Arn, Nit-ac, Puls,*	Nux-v, Sulph
Isolation		*Anac, Arg-n, Stram*	Plat, Puls
Jealousy	NUX-V, PULS	*Hyos, Ign, Lach, Phos*	Staph
Joy, excessive		*Acon, Puls*	Ars, Aur, Caust, Chin, Merc, Plat, Caust
Literary or scientific failure			Calc, Ign, Lyc, Nux-v, Puls, Sulph
Monetary losses	CALC-FL	*Arn, Ars, Ign,*	Aur, Calc, Nux-v, Puls
Position, loss of		*Ign, Plat*	Staph

MISCELLANEOUS Causative factors			
Pride, trauma-tized self-importance		*Calc, Lyc, Sulph*	Merc, Sil
Quarrels			Thuj
Rage, fury		*Arn, Aur*	
Rejection, abandonment, desertion, (compare to *Isolation*)	AUR, PSOR, PULS	*Arg-n, Cycl, Lach, Merc, Plat, Stram*	Anac, Bar-c, Calc, Cann-i, Carc, Chin, Kali-c, Sep
Reverses of fortune			Lach, Staph
Sexual celi-bacy, no sexual outlet		*Phos*	
Sexual ex-cesses	CALC, CHIN, LYC, NUX-V, PH-AC, PHOS, SEP, STAPH	*Kali-c, Merc*	Sil
Shame (See *Contradiction, Embarrass-ment, Humiliation, Indignation, Offense, Sex-ual abuse,*)			
Shock (See *Fright*)			
Surprises, pleasant			Chin, Merc

MISCELLANEOUS Causative factors			
Work, mental	STAPH, TUB	Anac, Arg-n, KaK c, Lach, Ph-ac, Sil	

SECTION 4.

FEARS AND PHOBIAS

Strong fears are good rubrics to use for arriving at the right
medicine for a psychoemotional case. Use the one felt strongest
by the patient, or having the highest reaction in a test, even if
there are several prominent fears. The others may resolve without treatment or can be treated later.

Accident, to a loved one			Ars, Caust
Agoraphobia	ACON, ARG-N, ARN	*Aur, Bar-c, Kali-ar, Lyc, Nat-m, Nux-v, Puls*	Am-m, Ars, Calc, Carb-an, Caust. Con, Ferr, Graph, Hep, Kali-bi, kali-c, Kali-p, Led, nat-a, Nat-c, Phos, Plb, Rhus-t, Sel, Stann, Sulph, Tab, Sil
Airplanes	ARG-N, CALC	*Acon*	Ars, Nat-m
Alone, being	ARG-N, ARS, HYOS, KALI-C, LYC, PHOS	*Puls, Sep, Stram*	
Alone, being, at night	STRAM	*Caust, Med*	Arg-n
Alone, being, lest he die	ARS	*Arg-n, Kali-c, Phos*	

Alone, being, and deliberately injuring himself			Ars, Merc, Sulph
Animals	BELL, CHIN, TUB	*Stram*	Calc, Carc, Caust, Hyos, Lyc, Med, Nat-m
Animals, imaginary	BELL		
Birds	IGN	*Nat-m*	
Blackness	STRAM		Ars, Tarent
Blind, becoming		*Nux-v, Sulph*	
Brilliant or shiny objects, mirrors, etc.			Cann-i, Lach, Stram
Burglars (looks under the bed for them)	ARS, NAT-M	*Arg-n, Bell, Ign, Lach, Lyc, Merc, Phos*	Anac, Aur, Sil, Sulph
Busy streets		*Acon*	Bar-c, Carc, Caust
Cancer	ARS, CALC, CARC, PHOS, PLAT	*Nit-ac*	Bar-c, Ign, Med, Nat-m, Sep
Cars, traveling in		Arg-n, *Aur, Lach, Sep*	Acon
Cats	DAPH	*Calc, Chin, Med, Tub*	
Cemeteries		*Stram*	
Church or theatre, when ready to go:	ARG-N		

Claustrophobia	ACON, ARG-N, LYC, NAT-M, PULS, STRAM	*Calc, Ign, Med*	Ambr, Ambr, Aran, Chin-a, Cocc, kali-ar, Nit-ac, Nux-v, Plb, Psor, Ruta, Staph, Succ, Sulph, Tab, Valer
Closed-in places	ACON, ARG-N, LYC, NAT-M, PULS, STRAM	*Calc, Ign, Med*	Ambr, Ambr, Aran, Chin-a, Cocc, kali-ar, Nit-ac, Nux-v, Plb, Psor, Ruta, Staph, Succ, Sulph, Tab, Valer
Company, desire to be alone	ANAC, BAR-C, CARB-AN, CHAM, CIC, GELS, IGN, NAT-M, NUX-V	*Ambr, Aur, Bel, Bry, Cact, Calc-p, Carb-v, Chin, Coloc, Cupr, Cycl, Ferr, Hell, Hep, Hyos, Iod, Lach, Led, Lyc, Nat-c, Plat, Puls, Rhus-t, Sel, Sep, Stann, Sulph, Thuj*	Acon, Alum, Anan, Ant-c, Ant-t, Bufo, Calc, Cann-i, Carb-s, Cedr, Cimic, Clem, Coca, Con, Cop, Dig, Dios, Elaps, Fl-ac, Graph, Grat, Ham, Helon, Hydr, Kali-bi, Kali-br, Kali-c, Kali-p, Kali-s, Mag-m, Mang, Nat-p, Nicc, Petr, Phos, Pic-ac, Psor, Sul-ac, Tarent, Til, Verat
Contagion	CALC, SULPH, LEUS	*Carc, Bor, Lach, Thuj*	Ars, Bufo, Bar-c, Sil

Control, losing		*Arg-n, Staph*	Cann-i, Carc, Ign, Med, Nat-m, Thea
Corners, walking past certain		*Arg-n*	
Creeping, something creeping out of every corner		*Phos*	Med
Crossing a bridge	ARG-N	Bor, Ferr	Bar-c, Lyc, Puls, Ter
Crowds, public places, agoraphobia	ACON, ARG-N, ARN	*Aur, Bar-c, Kali-ar, Lyc, Nat-m, Nux-v, Puls*	Am-m, Ars, Calc, Carb-an, Caust. Con, Ferr, Graph, Hep, Kali-bi, kali-c, Kali-p, Led, nat-a, Nat-c, Phos, Plb, Rhus-t, Sel, Stann, Sulph, Tab, Sil
Cruelty, upset by report of			Calc
Darkness	CANN-I, STRAM	*Acon, Calc, Lyc, Med, Phos, Puls, Sil*	Arg-n, Ars, Bell, Carc, Caust, Chin, Hyos, Nat-m, Sulph, Tub
Death	ACON, ARS, CALC, NIT-AC, PHOS, PLAT	*Arg-n, Am, Bell, Cann-i, Caust, Kali-c, Lach, Lyc, Med, Merc, Nat-m, Nux-v, Ph-ac, Puls*	

Death, when alone	ARS	*Arg-n, Arn, Kali-c, Phos*	Med
Dentist			Calc, Puls, Tub
Devil, being taken by the,			Anac, Lach, Manc
Disabled, be-coming	ARS		
Disasters		*Puls, Tub*	
Doctors	ACON, GELS, IOD	*Arg-n, Arn, Nux-v, Phos, Sep, Stram, Thuja, Tub*	Am, Ign, Nat-m, Verat-v
Dogs	BELL, CHIN, TUB	*Caust, Hyos, Puls, Stram,*	Calc, Carc, Lach, Med, Merc, Nat-m, Plat, Sep, Sil, Sulph
Driving			Calc, Scorp
Enemies			Anac, Hyos
Evil	CALC	*Arg-n, Ars, Caust, Chin, Lach, Nat-m, Phos, Sep, Staph, Stram*	
Failure			Arg-n, Am, Carc, Lyc, Nat-m, Phos, Sil, Sulph
Fainting		*Acon, Arg-n, Plat*	

Failing (See **Failure**)		*Stram*	Acon, Arg-n, Ars, Calc, Caust, Chin, Kali-c, Med, Nux-v, Phos, Sil, *Tub*
Flying	ARG-N, CALC	*Acon*	Ars, Nat-m
Germs (See **contagion**)			
Ghosts	LACH, PHOS	*Acon, Ars, Carbo-v, Caust, Hyos, Lyc, Manc, Med, Plat, Puls, Sep, Stram, Sulph*	Bell, Brom, Calc, Cann-i, Carc, Chin, Kali-c, Op, Sep, Spong, Thuj, Zinc
Happening, fear that something will happen	CAUST, NUX-V PHOS, PLAT, TUB	*Calc, Coloc, Nat-m, Ph-ac*	Ars
Happening, something terrible		*Ign, Sep*	
Heart disease		*Arn, Aur, Calc, Caust, Lach, Med, Phos*	Acon, Arg-n, Nat-m, Tarent
Heights (See **High places**)			
High places, vertigo	ARG-N	*Aur, Sulph*	Calc, Carc, Coca, Nat-m, Phos, Puls, Staph, Stram, Zinc
Humiliation		*Carc, Sep, Staph*	Lyc, Puls

Hurt, emotional	NAT-M, STAPH	*Arn, Ign*	Carc, Chin, Hep, Kali-c, Ruta, Spig
Imaginary things	BELL, STRAM	*Acon, Phos*	Ars, Lyc, Merc, Sep
Infection (See **Contagion**)			
Injury	STRAM		Arn, Ars, Aur, Cann-i, Chin, Hyos, Kali-c
Insanity	ANAC, CALC, CANN-I, PULS	*Med, Merc, Nat-m, Nux-v, Phos, Sep, Staph, Stram*	
Insects	CALC		Lyc, Nat-m, Phos, Puls, Sulph
Intruders will invade home	ARS, NAT-M	*Arg-n, Bell, Ign, Lach, Lyc, Merc, Phos*	Anac, Aur, Sil, Sulph
Invaded, home is	ARS, NAT-M	*Arg-n, Bell, Ign, Lach, Lyc, Merc, Phos*	Anac, Aur, Sil, Sulph
Job, losing			Calc, Ign, Puls, Sep, Staph, Sulph
Knives (keeps them out of sight)			Ars, Chin, Hyos, Merc, Nux-v
Late, being		*Arg-n*	Med
Lightning	PHOS, STAPH	*Calc, Coloc, Merc, Nat-m, Nit-ac, Sep,* Tub	Bell, Carc, Caust, Lach, Lyc, Sil, Stram, Sulph

Looked at, being	ARS	*Cham, Chin, Merc, Nat-m, Tub*	Bar-c, Calc, Nux-v, Sil, Stram, Sulph, Tarent, Thuj
Men		*Aur, Bar-c, Lyc, Merc, Nat-m, Plat, Puls*	Acon, Anac, Bell, Ign, Lach, Phos, Sep, Sulph
Music		*Acon*	Bar-c, Nit-ac, Nux-v, Phos, Sulph, Tarent, Thuj
Noise, in general		*Aur, Bell, Caust, Cham, Lyc, Med, Phos, Sil*	
Noise, from rushing water	STRAM	Hyos, Sulph	
Noise, at night		*Caust*	Bar-c
Noise, at the door		*Aur, Lyc*	
Noise, from street		*Caust*	Bar-c
Observed, one's condition being	CALC		
Ordeals		*Arg-n*	Arn, Ars, Thuj

People	HYOS, LYC, NAT-C, RHUS-T	*Acon, Anac, Aur, Bar-c, Carb-v, Caust, Iod, Kali-ar, Kali-c, Nat-m, Plat, Puls, Sep, Staph*	Alum, Ambr, am-m, Ars, Ars-I, Bell, Calc, Carb-an, Carb-s, Cupr, Dios, Ferr, Graph, Hep, Ign, Lach, Merc, Phos, Sel, Sep, Stann, Sulph, Tab, Til
People (in children)	BAR-C	*Lyc*	
Periods, during	IGN	*Lach, Nat-m, Ph-ac*	Acon, Bell, Nux-v, Phos, Plat, Staph, Sulph
Pins, sharp things	SIL	Ars, Merc, Nat-m, Plat	
Pointed objects	SIL, SPIG		
Poisoned, being (a specific form of suspicion)		*Ars, Bell, Hyos, Ign, Lach*	Anac, Nat-m, Ph-ac, Phos
Poverty, worry over money	ARS, CALC-FL	*Calc, Sep*	Kali-c, Nux-v, Puls, Staph, Sulph
Public, appearing in	GELS, SIL	*Carb-v, Lyc*	Anac, Arg-n, Plb
Punishment		*Carc, Ign, Staph*	Nat-m, Tarent
Self-control, losing		*Arg-n, Merc, Staph*	Cann-i, Nux-v, Sulph, Thuj
Shadows		*Calc*	Phos, Staph

Sleep, going to		*Lach*	Calc, Merc, Nat-m, Nux-v.
Snakes	LACH	*Bell, Carc*	Arg-n, Nat-m, Puls, Tub
Someone behind them	MED		Anac, Lach, Merc, Staph
Spiders		*Tarent*	Abel, Calc, Carc, Nat-m, Stram
Stage fright	ARG-N, ARS, CALC, CARC, IGN, LYC, MED, PHOS, PULS, SIL	*Anac, Bar-c, Caust, Ph-ac,*	Nat-m
Stomach, "butterflies" in	ARS	*Aur, Bar-c, Chin, Kali-c, Sulph, Tarent*	
Strangers		*Bar-c, Thuj*	Carc, Caust, Lach, Lyc, Puls, Sil, Stram, Tub
Streets, busy		*Acon*	Carc, Caust
Strangled, being	PLAT		
Struck, being, by people coming towards him or her	ARN		Bell, Ign, Kali-c, Lach, Stram, Thuj
Subways, underground	ACON, STRAM		
Suffocation	ACON	*Ars, Phos, Staph, Stram, Sulph*	Lach, Merc, Nux-v

Suffocation, at night		*Ars, Chin, Lyc, Med, Puls, Sil, Suph*	Arn
Suicide		*Ars, Merc, Nux-v*	Arg-n, Lach, Med, Plat, Sep, Tub
Thunderstorms	PHOS, STAPH	*Calc, Coloc, Merc, Nat-m, Nit-ac, Sep,* Tub	Bell, Carc, Caust, Lach, Lyc, Sil, Stram, Sulph
Touch		*Acon, Arn, Bell, Cham, Chin, Kali-c, Nux-v, Tarent*	
Tunnels	STRAM		Acon, Arg-n, Bell, Nat-m
Undertaking anything		*Arg-n, Ars, Lyc*	Nux-v, Sil
Violence (See **Burglars, Injury, Struck**)			
Washing obsessively	SIL, SYPH	*Sulph*	
Water (See also: **Noise, from rushing water**)	HYOS, STRAM	*Lach, Phos*	Bell, Cann-i, Carc, Med, Merc, Nux-v, Sulph, Tarent
Wind	CHAM	*Thuj*	
Women			Puls, Sep, Staph
Work		*Arg-n, Kali-c, Nux-v, Puls, Sil, Sulph*	Calc, Cham, Chin, Coloc, Hyos, Lyc, Nat-m, Phos

SECTION 5.
DELUSIONS, ILLUSIONS, IDEATIONS

Affection of friends, has lost	AUR		Hura
Animals, delusions of	OP	*Ars, Bel, Calc, Cin, Crot h, Hyos, Stram*	Absin, Aeth, Cham, Cina, Colch, Con, Lac-c, Lyss, Mad, Puls, Sec, Sulph, Tarent, Thuj, Valer
Appreciated, that she is not		*Pall*	Plat
Arrested, is about to be		*Zinc*	Arn, Ars, Bel, Cupr, Kali dr, Meli, Plb
Beautiful, things seem (even if not)	SULPH	*Cann-i, Lach*	Bel, Coca
Blames self		*Acon, Ars, Aur, Dig, Hyos, Ign, Nat-m, Op, Puls, Sarr, Thuj*	Calc-p, Cob, Cycl, Hell, Hura, Lyc, Merc, Nat-a, Ph-ac
Body, delusional ideas about the state of	SABAD	*Cann-I, Hyos*	Alum, Anac, Kali-c, Op, Pic-ac, Stram
Body, delusion of superiority of		*Cann-i, Plat*	Staph
Cancer, delusion of having	CARC	*Verat*	

Crime, as if he had committed		*Ign, Merc,*	Alum, Anac, Carb v, Cob, Cycl, Dig, Hyos, Kali-br, Op, Phos, Sabad, Thuj
Criticized, that she is		*Bar-c*	Laur, Plb, Rhus r
Dead persons, sees		*Anac, Ars, Hep, Hyos, Kali-br, Kali-c, Lach, Mag c, Ph-ac, Phos, Plat*	Agar, Alum, Amm c, Arg-n, Arn, Ars i, BAR-C, Bel, Brom, Bry, Calc, Camth, Caust, Cocc, Con, Fl ac, Graph, Hura, Iod, Kali-ar, Kali p, Laur, Mag m, Nat c, Nat n, Nat p, Nit-ac, Nux-v, Op, Plb, Ran sc, Sars, Sil, Strych, Sulph, Sul ac, Thuj, Verb, Zinc
Deserted, forsaken	ARG-N	*Cycl, Kali-br, Plat, Stram*	Bar-c, Camph, Cann-i, Carb an, Carb v, Chin, Hura, Hyos, Lil t, Lyss, Nat c, Pall, Puls

Demons, devils, is possessed by			Hyos
Demons, devils, sees		*Anac, Bel, Hell, Plat, Puls, Zinc*	Ambr, Ars, Cann-i, Cupr, Dulc, Hyos, Kali-c, Lach, Nat c, Op, Stram, Sulph, Zinc
Die, believes about to	ACON	*Arg-n, Chel, Croc, Nit-ac, Thuj*	Bar-c, Cann-i, Cupr, Kali-c, Lac d, Nux-v, Petr, Podo, Rhus t, Stram
Disease, that he has every		*Aur m*	Stram
Disease, has incurable		*Arg-n, Sabad*	Cact, Chel, Plb, Verat
Distances (see **Enlarged**)			
Divided into two, being double		*Anac, Bapt, Nux m, Petr, Stram*	Alum, Cann-i, Glon, Lach, Lil t, Mosch, Puls, Sil, Thuj
Enlarged, de-lusions of being	CANN-I	*Op, Plat*	Acon, Alum, Bel, Berb, Coc c, Euph, Glon, Hyos, Laur, Nat c, Nux-v, Pic ac, Sabad, Stram, Zinc

Enlarged, distances seem	CANN-I		Camph, Camn s, Glon, Nux m, Stann
Fire, delusions of		*Bel, Calc, Hep, Puls*	Alum, Amm m, Anac, Ant-t, Ars, Cal p, Clem, Croc, Daph, Kali n, Kros, Laur, Lyss, Mag m, Nat-m, Phos, Plat, Rhod, Rhus t, Spig, Spong, Stann, Stram, Sulph, Zinc
Floating in air		*Lach, Nux m*	Asar, Bel, Cann-i, Canth, Hura, Kali-br, Phos
Flying, sensation of		*Cann-i*	Asar, Camph, Lach, Oena, Op
Friend, thinks has lost the affection of (see *Affection*)			

Ghosts, sees	BEL	*Camph, Cupr, Hyos, Nat-c, Nat-m, Stram, Sulph*	Agar, Alum, Ambr, Amm-c, Ant-t, Ars, Aur, Bov, Carb-v, Cocc, Dulc, Hell, Hep, Hura, Hyper, Ign, Kali-c, Lach, Lcy, Merc, Nit-ac, Op, Phys, Plat, Puls, Sars, Sep, Sil, Spig, Tarent, Thuj, Zinc
God, is being punished by			Kali-br
God, is com-municating with			Stram, Verat
Grandeur, de-lusions of		*Agar, Cann-i, Plat, Staph*	Aeth, Bel, Cupr, Lyss, Phos, Sulph, Verat
Hallucinations, visual	CANN-I, BEL	*Calc-s, Cann-s, Carb-s, Crot-c, Hep, Hyos, Lach, Nat-m, Op, Puls, Sil, Stram, Sulph*	Absin, Alum, Ambr, Arg-n, Calc, Canth, Cench, Cham, Cic, Cimic, Con, Graph, Lyc, Nit-ac, Nux-m, Plat, Rhod, Rhus-t, Sep, Spong, Va-ler

Idealization (See **Beautiful, things seem**)			
Identity, errors of personal			Alum, Ant-c, Bapt, Cann s, Gels, Lach, Petr, Phos, Plb, Stram, Thuj, Valer
Insane, that she will become	CIMIIC	*Acon, Chel, Manc*	Calc, Cann-i, Lil t, Med, Merc, Nat-m, Tanac
Insects, sees	ARS, BEL	*Lac-c, Stram*	Caust, Dig, Hyos, Merc, Phos, Plb, Puls, Tarent
Intruders are invading home	ARS, NAT-M	*Arg-n, Bell, Ign, Lach, Lyc, Merc, Phos*	Anac, Aur, Sil, Sulph
Invaded, home is	ARS, NAT-M	*Arg-n, Bell, Ign, Lach, Lyc, Merc, Phos*	Anac, Aur, Sil, Sulph
Kleptomania		*Absin, Art-v, Cur, Nux-c*	Ars, Bry, Caust, Kali-c, Lyc, Puls, Sep, Staph, Stram, Sulph, Tarent
Large, parts of body seem to be		*Hyos*	Alum, Op, Pic-ac, Stram

349

Laughed at, that he is		*Bar-c*	
Looking at her, everyone is			Meli, Rhus-t
Murder, thinks she will murder her family			Jab, Kali-br
Murdered, that he will be		*Bel, Calc, Hyos, Op, Rhus-t, Stram*	Amm-m, Ars, Ign, Kali-c, Lact, Lyc, Mag-c, Merc, Phos, Plb, Verat, Zinc
Music, thinks he hears	CANN-I	*Lach, Stram*	Croc, Lyc, Plb, Puls, Sal-ac, Thuj
People, sees		*Ars, Bel, Bry, Hyos, Puls, Stram*	Chin, Con, Kali-c, Lyc, Lyss, Mag-s, Med, Nat-m, Op, Plb, Rheum, Sep, Sulph, Thuj, Valer, Verat
Persecuted, being	CHIN, DROS	*Bell, Hyos, Kali-br, Lach, Puls*	Absin, Anac, Ars, Aur, Cic, Con, Crot-h, Cupr, Hell, Lys, Meli, Merc, Nat-c, Plb, Rhus-t, Sil, Stram, Strych, Zinc

Person is behind him			Anac, Brom, Calc, Casc, Cench, Crot-c, Lach, Med, Ruta, Staph, Sil
Person is beside him		*Ars*	Anac, Apis, Bell, Calc, Camph, Carb-v, Cench, Nux-v, Petr, Thuj, Valer
Person is also in the room			Cann-i
Places, that he is in two at the same time		*Cench, Lyc, Sil*	
Pursued by enemies		*Bell, Chin, Hyos, Lach, Puls*	Absin, Anac, Ars, Aur, Cic, Con, Crot-h, Cupr, Dros, Hell, Kali-br, Lys, Meli, Merc, Nat-c, Plb, Rhus-t, Sil, Stram, Strych, Zinc
Smaller, sensation of being			Acon, Agar, Calc, Carb, Sabad, Tarent
Snakes, around him		*Hyos, Lac-c*	Arg-n, Bel, Calc, Lach, Op
Spiders, sees		*Lac-c*	

Strange, everything seems		*Graph, Nux-m, Plat*	Bar-m, Carb-an, Cic, Staph, Stram
Thieves, sees		*Ars, Nat-m*	Alum, Aur, Bel, Cupr, Kali-c, Mag-c, Mag-m, Merc, Nat-c, Petr, Phos, Sil, Verat, Zinc
Time, exaggeration of	CANN-I	*Cann-s*	Nux-m, Onos
Voices, hears		*Cham, Coff, Crot-c, Elaps, Kali-br, Phos*	Abrot, Agar, Anac, Aster, Bel, Benz-ac, Cann-i, Carb-s, Cench, Coca, Crot-h, Hyos, Lac-c, Lach, Lyc, Manc, Med, Nat-m, Petr, Plb, Stram
Watched, that he is being	ARS	*Bar-c, Hyos*	Rhus-t

SECTION 6.
SLEEP DISTURBANCES

Sleepless from mental activity	ARS, CALC, COFF, HEP, NUX-V, PULS	*Ambr, Arg-n, Bry, Carb-s, Chin, Cocc, Ferr, Fl-ac, Gels, Graph, Hyos, Kali-c, Lach, Lyc, Nat-m, Psor, Sep, Sil, Staph, Sulph, Tub, Zinc*	Agar, Alum, Bar-c, Bell, Bor, Cact, Caust, Con, Cupr, Hell, Hyper, Ign, Nat-p, Plb, Spig, Thuj, Viol-o
Waking, frequent	ALUM, BAR-C, CALC, HEP, MUR-AC, PHOS, PULS, SEP, SULPH	*Acon, Ambr, Ant-c, Arg-n, Ars, Bar-m, Bell, Bor, Bry, Calc-ar, Carb-an, Carb-s, Carb-v, Card-m, Caust, Chin, Coff, Dig, Dros, Graph, Kali-c, Lach, Lyc, Mag-c, Merc, Nat-m, Nit-ac, Nux-v, Plat, Ran-b, Rhus-t, Sil, Spig, Stann, Staph, Stront, Zinc*	Agar, Agn, Am-C, Am-M, Anac, Ant-T, Apis, Arn, Aur, Benz-Ac, Berb, Bufo, Cann-S, Canth, Caps, Cham, Chel, Chin-A, Cimic, Cocc, Colch, Coloc, Con, Cupr, Cycl, Dulc, Ferr-I, Fl-Ac, Guaj, Hura, Hyos, Ign, Ip, Kali-Ar, Kali-S, Mag-M, Mag-S, Mang, Merc-C, Mez, Mosch, Naja, Nat-A, Nat-C, Nat-

Waking, frequent, cont'd.			P, Nat-S, Nux-M, Ox-Ac, Petr, Ph-Ac, Rhod, Ruta, Sabad, Sabin, Sang, Sars, Sel, Sene, Seneg, Spong, Stram, Sul-Ac, Tarax, Ter, Thea, Thuj, Vert, Viol-O, Viol-T
Dreams, frightening	AMM-M, ARS, AUR, BOR, CALC, CALC-AR, CARB-V, CINA, COCC, CON, CROT-C, GRAPH, KALI-AR, KALI-BR, KALI-C, LYC, NAT-M, NICC, PULS, SIL	*Ant-C, Arg-M, Arn, Bapt, Bell, Bry, Calc-F, Calc-P, Calc-S, Camph, Cann-S, Cham, Chel, Chin, Colch, Cycl, Eup-Pur, Hyos, Kali-i, Kali-S, Laur, Mag-C, Mag-M, Med, Merc, Nat-A, Nat-C, Nat-S, Nit-Ac, Nux-V, Phos, Sulph, Zinc*	Acon, Agar, Agn, Alumn, Am-C, Apis, Aur-M, Bar-C, Bar-M, Bufo, Cann-i, Carb-An, Carb-S, Caust, Chin-A, Chin-S, Cimic, Clem, Coloc, Cop, Dig, Dios, Dros, Dulc, Elaps, Euphr, Fl-Ac, Guaj, Hep, Hydr, Hyper, Ign, Ip, Iris, Kali-Bi, Kali-i, Kali-S, Lil-T, Lyss, Mang, Merc-C, Mez, Mosch, Mur-Ac,

Dreams, frightening, cont'd.			Naja, Nux-M, Op, Ox-Ac, Petr, Phel, Ph-Ac, Plat, Psor, Rhus-T, Sa-bad, Sang, Sars, Sec, Sep, Spig, Stann, Staph, Stram, Stront, Tab, Thea, Thuj, Ust, Verb, Verat, Verat-V, Zing

Bibliography

- Boericke, W., and Boericke, O., *Pocket Manual of Homeopathic Materia Medica With Repertory,* Boericke & Runyon, 1927
- Chappell, P., *Emotional Healing with Homoeopathy*, 1994 Element Books, Ltd., England
- Gallavardin, J., *Repertory of Psychic Medicines with Materia Medica*, B. Jain, Pub.
- Kent, J.T., *Repertory of the Homeopathic Materia Medica,* 5th Edition, Ehrhart & Karl, 1945
- Murphy, R., *Homeopathic Medical Repertory*; Hahnemann Academy of North America, 1993
- Smith, T., *Homeopathic Medicine for Mental Health,* Healing Arts Press 1989

www.ingramcontent.com/pod-product-compliance
Lightning Source LLC
Chambersburg PA
CBHW021548210326
41599CB00010B/363